M^2 Models and Methodologies for Community Engagement

Reena Tiwari • Marina Lommerse • Dianne Smith
Editors

M^2 Models and Methodologies for Community Engagement

Springer

Editors
Reena Tiwari
Faculty of Built Environment
Curtin University
Perth, Western Australia
Australia

Marina Lommerse
Marina Lommerse Consultants
South Fremantle, Western Australia
Australia

Dianne Smith
Faculty of Built Environment
Curtin University
Perth, Western Australia
Australia

ISBN 978-981-10-1179-5 ISBN 978-981-4585-11-8 (eBook)
DOI 10.1007/978-981-4585-11-8
Springer Singapore Heidelberg New York Dordrecht London

Printed on acid-free paper

Springer is part of Springer Science+Business Media (www.springer.com)

Foreword

Since the 1970s, the role of science and academics in general and of social science as self-reflexive and critical has come under increasing scrutiny. Critical research has aimed at a new progressive relationship between action and reflection, theory and practice. There has been a strong concern with awareness-raising and empowerment of hitherto marginalized and oppressed groups. Interesting precursors are Kurt Lewin (action research), Paulo Freire (awareness-raising) and Robert Jungk (Futures Workshop). After founding a Centre Paulo Freire in Vienna, we have been involved in diverse partnerships with local schools, green and anti-globalization activists, as well as setting up the first social platform financed by the European Commission, with the objective of fostering social cohesion in the city.

In a world dominated by neoliberal thinking, we are suffering from increasing commercialization and the subordination of academics as well as formal political power-holders to the needs and interests of economic power. Although there is a megatrend of fostering researchers to become 'relevant' and market-prone, promising alternatives have been flourishing at the margins and interstices. To me, *M^2 Models and Methodologies for Community Engagement* is an exciting testimony to the richness of community engagement as one example of a broad, worldwide, bottom-up movement which aims at making social science matter and work for a better world. Different case studies in this collection show the potential of a dialogue between transdisciplinary theories and practices in forms of community development that not only empower the community, but are rewarding for the professionals as well.

From the diverse learning and research partnerships in which I have had the pleasure of participating, I think that three main lessons emerge for transdisciplinary cooperation. First, researchers tend to neglect the importance of organizational issues in transdisciplinary research. Even researchers who sympathize with interdisciplinarity and recognize the importance of relevant research may lack the practical tools to make their knowledge matter outside the academy. Therefore, I very much appreciate that this book is not only dedicated to methods and tools, but also to models of practice involvement on a practical and organizational level. In our research, we have observed that good learning and research partnerships require trust and mutual understanding. We have coined the term *knowledge alliance* to describe a reality, as well as an aspiration to lasting, respectful partnerships of knowledge production, diffusion and mobilization.

Secondly, I appreciate the importance given to power and empowerment in this collection. Researchers and practitioners will be reminded and encouraged to take the diverse dimensions of power relations into account when they read the case studies in this collection, which exemplify a multi-faceted awareness of and respect for the complexity of power relations in any community. Power can be fluid and all-pervasive in the Foucauldian sense, but also very concrete and personalized. All too often, empowerment implies that someone in power—who coincidentally often finances research—loses part of their power. Making decisions which empower the community in question requires an intellectual and political effort to appraise and balance social, political and economic forces.

Thirdly, creating happy communities gives great personal rewards to social scientists, as much as to the communities they work with. But only rarely is this recognized academically. One of the key merits of books like this one is its potential to overcome this duality and demonstrate that good transdisciplinary research not only leads to happier communities, but also to better knowledge about community dynamics for future practice. Transdisciplinarity has the key advantage over conventional research that it can mobilize the local knowledge of stakeholders systematically. A scientific approach which links these contextualized insights to existing social theories, as in this collection, will benefit social science, its practitioners and the communities served, and could pave the way towards a governance of knowledge beyond neoliberalism.

Andreas Novy
Chairperson of the Austrian Green Foundation
Associate Professor, Urban and Regional Development,
Institute for the Environment and Regional Development,
Vienna University of Economics and Business

Two roads diverged in a wood, and I—
I took the one less traveled by,
And that has made all the difference.
Robert Frost, *Mountain Interval*

Preface

The book was conceived as a project of the International Cooperation Research Cluster (ICRC), established in early 2012 at Curtin University. The cluster formation was a University initiative to bring together researchers, practitioners and postgraduate students from various disciplines working in similar fields, with the objective of building each other's capacity by investigating community development and cooperation projects from a research perspective.

Synergies were established following an exercise of mapping individual cluster members' research strengths, the nature of their work and the location of these works globally. Community-centric projects based in different disciplines and in various international contexts emerged as common ground for further exploration by the ICRC cluster members.

Questions for exploration were:

- How do researchers and professionals from different disciplinary perspectives working on community-based projects of different nature and scales and at different locations in the developing and developed world, engage with their own communities of practice and the communities under study?
- What are the effects of these engagements for both practitioners and local community members? Is capacity built in any form for the participants?
- If so, what are the models and methodologies of these engagements? What are the methods employed in these interactions with community?

An idea of a book dealing with the above questions was formed and *M² Models and Methodologies for Community Engagement* was born. The book unfolded as an act of co-production and co-creation, where academics, researchers, practitioners, graphic artists and editors were involved in an active process of threading together the myriad voices (personal and collective) and vignettes of perspectives.

Traditionally, a contributed book is structured as a series of stand-alone chapters on a targeted subject, bookended by a contextualizing introduction and a concluding chapter by the editor(s). Typically, the contributed chapters are a result of selected conference papers on a theme, or a call for interest, or an invitation to experts in the field to write on a targeted subject set out by the publisher/editor(s). Authors most often have no direct contact with each other, and therefore do not impact on the development of the others. The evolution of this book was quite different, with authors contributing from the beginning towards setting the direction for the book. Writing workshops were organized and brainstorming sessions between the authors, followed by editorial sessions, demonstrate a shared creativity, a collective efficacy. Parts II and III evolved out of the workshops sessions once Part I of the Book was developed. For us, this book, an object of co-creation, helped us in opening up to new perspectives, embracing novel ways of moving forward and building on each others' works. The process aided us on a path less travelled and brought the book contributors together as an active, engaging, non-placed global community.

The editors would like to thank and acknowledge the following people for their creative participation, and for igniting debate and development of *M² Models and Methodologies for Community Engagement.*

- The contributing authors for their enthusiastic response to the intention of the book, and their ideas that shaped the book through the workshops, questionnaires and contributed chapters.
- The invited reviewers, all of Curtin University, Western Australia, for considered and constructive assessment of the editorial chapters, the book structure and participation in writing workshops:
  - Bob Pokrant, School of Media, Culture & Creative Arts;
  - Jaya Earnest, Director of Graduate Studies, Health Sciences, and Associate Professor of International Health in the Centre for International Health;
  - Mohammed Ali and Clancy Read, Centre for International Health;
  - Sarah McGann and John Stephens, School of the Built Environment.
- Andreas Novy, Chairperson of the Austrian Green Foundation, and Associate Professor for Urban and Regional Development at RUW (Institute for the Environment and Regional Development) at WU Vienna, for writing the foreword.
- Christina Houen, *M2*'s Production Coordinator, Adjunct Research Associate of School of Media, Culture and Creative Arts, Curtin University, and Editor/Director of www.perfectwordsediting.com, for her dedicated editing and production work, impressive attention to detail and timely advice to authors.
- Allyson Crimp, Designer, of et-al.com.au, for her graphic work and image editing.
- Jessica Winters, planning student, and Robyn Creagh, PhD candidate, Curtin University, for their research assistance, dedication and contributions to the book preparation.

We thank Springer for their enthusiastic approach to publishing the book.

Last but not least, we thank Curtin University for support enabling the research and production associated with this book project.

Contents

Contributors

Mohammed Ali Centre for International Health, Curtin University, Perth, Australia

Denise K. DeLuca BCI: Biomimicry for Creative Innovation, Seattle, USA

Salvatore Di Mauro Queensland College of Art, Griffith University, Brisbane, Australia

Jaya Earnest International Health, School of Nursing and Midwifery, Curtin University, Perth, Australia

Centre for International Health, Curtin University, Perth, Australia

Anne Farren Department of Design, Curtin University, Perth, Australia

Mark Jones International Health, School of Nursing and Midwifery, Curtin University, Perth, Australia

May Lample Kimanya-Ngeyo Foundation for Science and Education, Jinja, Uganda

Marina Lommerse Marina Lommerse Consultants, South Fremantle, Australia

School of the Built Environment, Curtin University, Perth, Australia

Damian Madigan School of Art, Architecture and Design, The University of South Australia, Adelaide, Australia

Sarah McGann Department of Architecture and Interior Architecture, Curtin University, Perth, Australia

Carmen Mendoza-Arroyo Department of Urban Design and Regional Planning, School of Architecture, Universitat Internacional de Catalunya, Barcelona, Spain

Barbara Milech School of Communication, Culture and Creative Arts, Curtin University, Perth, Australia

David Morris School of Art, Architecture and Design, The University of South Australia, Adelaide, Australia

Yatin Pandya FOOTPRINTS Environment Architecture Research Technology Housing (E.A.R.T.H), Ahmedabad, India

Bob Pokrant Department of Social Science and International Studies, Curtin University, Perth, Australia

Veena Poonacha Research Center for Women's Studies, Dr. Avabai Wadia Archives, SNDT Women's University, Mumbai, India

Mokhlesur Rahman Curtin University, Perth, Australia

Center for Natural Resource Studies, Dhaka, Bangladesh

Clancy Read Centre for International Health, Curtin University, Perth, Australia

Centre for International Health, Curtin University, Perth, Australia

Debra Singh Kimanya-Ngeyo Foundation for Science and Education, Jinja, Uganda

Dianne Smith Department of Architecture and Interior Architecture, Curtin University, Perth, Australia

Nancy Spanbroek Department of Architecture and Interior Architecture, Curtin University, Perth, Australia

Reena Tiwari Departments of Urban and Regional Planning and of Architecture and Interior Architecture, Curtin University, Perth, Australia

Pere Vall-Casas Department of Urban Design and Regional Planning, School of Architecture, Universitat Internacional de Catalunya, Barcelona, Spain

About the Editors

Reena Tiwari is Associate Professor in the Departments of Urban & Regional Planning and Architecture at Curtin University, and Director of the International Cooperation Research Cluster. Her work focuses on the integration of urban development and place-making, with an emphasis on the importance of community involvement. Reena has received an Australian National Award for Teaching Excellence for leading and inspiring a model of teaching and research which has established a point of difference in the profession, in its commitment to community engagement.

Marina Lommerse is a designer, educator, curator, researcher and strategist with Marina Lommerse Consultants, and Adjunct Professor, Curtin University. She works with creative people and communities to strengthen individual and collective capacity and build supportive environments, mobilizing creative projects that promote positive change in communities, through publications, creative works, and curatorship of exhibitions.

Dianne Smith is Associate Professor and Head of Program in Interior Architecture at Curtin University, and Adjunct Professor at Queensland University of Technology. Dianne's research focuses on design for the cognitively impaired, the meaning of environments, the impact of colour on experience and our understanding of place, and design education.

About the Editors

About the Authors

Mohammed Ali coordinates postgraduate coursework programs in International Health at Curtin University. He worked as a physician-researcher in diarrhoeal diseases and nutrition in Bangladesh, and has conducted research in cardiac rehabilitation and forensic mental health services among Australian Indigenous people.

Denise DeLuca is Director and Co-Founder of BCI: Biomimicry for Creative Innovation, an emerging network that shares nature's principles with businesses and organizations to achieve radical transformation. Denise is also an adjunct faculty in the Sustainable Design Program at the Minneapolis College of Art and Design, and a biomimetic consultant for Swedish Biomimetics 3000.

Salvatore di Mauro is a Senior Lecturer in Design at Griffith University. Since 1997, he has focused less on his artwork in a gallery context and more on working with communities to develop public art and design needs. He finds working with creative thinkers, artists, crafts persons, designers and local industry both challenging and rewarding.

Jaya Earnest is Director of Graduate Studies in the Faculty of Health Sciences and Associate Professor of International Health at the Centre for International Health, Curtin University. She has worked for 26 years in India, Kenya, Uganda, Rwanda, East Timor and Australia as an educator and researcher. Her central research interests are post-conflict adversity, resilience among refugee and migrant populations and interdisciplinary cross-cultural participatory research.

Anne Farren is Academic Coordinator of Fashion and Director of the Fashion Design & Research HUB at Curtin University. She maintains a creative practice as both a maker and curator of fashion and textiles. Her research interests include investigation into the relationship between design practice and technology.

Mark Jones is Professor for transcultural health improvement at Curtin, and Director of the Western Australian Global Health Alliance. Mark has built on his background in primary health care nursing, health policy development, and international nursing expertise to make a significant contribution to improving capacity and capability within health care systems of developing countries.

May Lample works as a research assistant at Kimanya-Ngeyo Foundation for Science and Education in Jinja, Uganda. She is also researching for her Master's degree in Public Health, majoring in Maternal and Child Health at University of California (Berkeley) School of Public Health.

Damian Madigan is a Lecturer in Architecture at the University of South Australia, Director of Madigan Architecture and a PhD candidate in the School of Art Design & Architecture at Monash University, where he is researching models of flexible housing.

Carmen Mendoza-Arroyo is Professor at the Universitat Internacional de Catalunya, and Co-Principal of the firm DACarquitectura. She has recently been appointed Sub-Director of the School of Architecture at UIC. Her research focuses on sustainable urban regeneration and development.

Sarah McGann is a Senior Lecturer in the Department of Architecture and Interior Architecture at Curtin University. Her particular interest is in the intersection of architecture and social theory. Sarah's latest publication, *The Production of Hospice Space,* draws on her practice designing health, hospitality and recreation projects.

Barbara Milech Adjunct Professor, served as Curtin University's Humanities Director of Graduate Studies for more than a decade; here she worked to establish frameworks that support creative-production research. Her supervision and research interests are in literature and creative writing, and draw on narrative, cultural, feminist and Lacanian studies.

David Morris is an architect, builder and director of the Student Construction Program in the School of Art, Architecture and Design at the University of South Australia. The program has constructed many award winning projects, which focus on culturally responsive designs for Indigenous communities. David Morris is recipient of the Australian National Award for Programs that Enhance Learning for the Vanuatu Project discussed in the book.

Yatin Pandya is an author, activist, academician, researcher and practising architect in his firm FOOTPRINTS E.A.R.T.H. (Environment Architecture Research Technology Housing) at Ahmedabad. Environmental sustainability, socio-cultural appropriateness, timeless aesthetics and economic affordability are key principles of his work.

Bob Pokrant is Professor of Anthropology at Curtin University, Australia. His main research interests are fisheries, aquaculture, and community adaptation to climate change in South Asia, particularly Bangladesh.

Veena Poonacha is Professor and Director of the Research Centre for Women's Studies at SNDT Women's University in Mumbai, India. She is also honorary director of the Centre for Rural Development that runs empowerment and development projects for rural women in Southern Gujarat in Western India.

Mokhlesur Rahman is a development practitioner in the fields of community based natural resource management in wetlands, forests and coastal zones, and community livelihoods, climate change adaptation and disaster risk reduction in Bangladesh.

Clancy Read completed her Doctor of Philosophy in International Health in the Centre of International Health (CIH), Curtin University. She is currently cofounder and lead consultant of I.D.E.A. Global Consulting offering research, project management and training services for humanitarian organizations.

Debra Singh is a medical doctor with a Masters in Public Health and is currently a PhD candidate at Sydney University. She and her husband established the Kimanya-Ngeyo Foundation for Science and Education in Jinja, Uganda. Her research and education focus is on female and youth empowerment through participation in health care processes and community health worker training.

Nancy Spanbroek is a Senior Lecturer in Architecture Interior Architecture at Curtin University. Her design research explores the interface between design pedagogy, community and exhibition as a testing ground for creative practice.

Pere Vall-Casas is Associate Professor at Escola Tècnica Superior d'Arquitectura de la *Universitat International de Catalunya*. His research focuses on cultural landscapes. He is currently working on urban sprawl repair through research into pre-urban patterns and participant observation.

List of Figures

List of Tables

Chapter 1
Navigating Community Engagement

Dianne Smith, Reena Tiwari and Marina Lommerse

Abstract What it is about community involvement that attracts some professionals to adopt ways of working that embrace the community members as partners? Which aspects make community work rewarding for a professional, and more importantly, successful from a community member's perspective? The theoretical constructs—community engagement, capacity building, and community empowerment—will be discussed in order to demonstrate how theory and practice are relevant to the development of ways to be involved in communities. A framework that we consider is of value has evolved that enables us to map or describe the attributes of community based projects; that is, an approach which aims to move beyond simply bringing people together from a variety of disciplines, to one which is transdisciplinary and applicable across cultures and genres of projects. Although a transdisciplinary approach is not new in itself, by making it explicit as an aspiration, we highlight the possible limitation of those projects that only bring together differing contributors at core moments for their expertise, without reflecting or planning for the potentially new ways of conceptualizing and of actioning what needs to be done. Such interactions are discussed in relation to participation and engagement. By constructing a project as transdisciplinary, all people—including the community—are ongoing contributors, who are able to wander into others' discipline-specific arenas and vice versa.

D. Smith (✉)
Department of Architecture and Interior Architecture, Curtin University, Perth, Australia
e-mail: dianne.smith@curtin.edu.au

R. Tiwari
Departments of Urban and Regional Planning and of Architecture and Interior Architecture, Curtin University, Perth, Australia
e-mail: r.tiwari@curtin.edu.au

M. Lommerse
Marina Lommerse Consultants, South Fremantle, Australia

School of the Built Environment, Curtin University, Perth, Australia
e-mail: marina@marinalommerse.com

R. Tiwari et al. (eds.), *M² Models and Methodologies for Community Engagement,*
DOI 10.1007/978-981-4585-11-8_1, © Springer Science+Business Media Singapore 2014

Keywords Community engagement · Community participation · Transdisciplinary · Capacity building

Introduction

Navigating the quagmire of considerations around community-based projects commences way before one gets to the location in question. This book strives to assist those considering such an adventure, and/or reflecting upon past interactions, by providing a collection of insights and strategies gleaned from the contributing authors' many varied experiences. We hope that this exploration will inspire many to choose to engage in community-based projects which involve the community as active participants and partners; an experience that the editors have found to be both challenging and rewarding.

When professionals set out to engage with a community, they enter a complex field of possibilities. Firstly, there is a multitude of ways of selecting communities, as well as those for being chosen by or appointed to a community. Forms of interaction and modes of engagement may differ widely, depending on the scale and complexity of the issues, the work to be undertaken, and its locational context. The duration and intimacy of the project, ranging from those involving a single household or act to those that may cover an entire region or issue, can also add to the complexity of issues around the tasks involved.

Therefore, we aim to consider what it is about community involvement that attracts some professionals to working this way, and what makes some community work rewarding for the professional, and more importantly, successful from the community members' perspective. The theoretical constructs—community engagement and empowerment, and capacity building—will be discussed in order to demonstrate how theory and practice are relevant to the development of ways to be involved in communities. This also positions the subsequent chapters in a framework that we consider is of value; that is, an approach which aims to move beyond simply bringing people together from a variety of disciplines, to one which is transdisciplinary and applicable across cultures and genres of projects. Although a transdisciplinary approach is not new in itself, by making it explicit as an aspiration, we highlight the possible limitation of those projects that only bring together differing contributors at core moments for their expertise, without reflecting or planning for the potentially new ways of conceptualizing and of actioning what needs to be done. By constructing a project as transdisciplinary, all people are ongoing contributors, who are able to wander into others' discipline-specific arenas and vice versa. We will expand on this overarching approach after clarifying the objectives of community intervention: engagement and empowerment, and capacity building.

Community Engagement and Empowerment

Community engagement and participation is an approach that has been widely debated and has generated varied opinions regarding its relevance and value (for example, Hall 2010; Israel et al. 1998; Israel et al. 2001; Kumar et al. 2011; Minkler and Wallerstein 2008; Mooney 2005; Patel et al. 2011a; Tsey 2008; Tsey et al. 2009; Wallerstein and Duran 2006; Wallerstein and Duran 2010).

These practices have prevailed; even in 1993, Dudley recognised the increasing use of this participatory approach in generating public policy. However, concerns have continually been raised because of practices which claim to engage communities, yet lack the participation of all members of the community at all levels of the project. Participation by all the various groups may not yet be possible because it is difficult to gain clarity concerning purpose and actions which all will fully understand, and therefore be willing to engage in. For example, Kumar notes the basic terms 'community' and 'participation' are understood differently, depending on the context of the project, and therefore, potential participants may be comfortable with, or wary of, the opportunity to engage (Kumar 2005, p. 55). For those who have influential vested interests, community participation as a methodology may pose a potential threat (Dudley 1993).

However, as Dudley notes, community participation can be used by professionals as both a tool and a goal. If used as a tool, it has the risk of creating an obvious 'us and them' divide, whereas, when used as a goal, it has the potential to transfer power to the community more equitably (Dudley 1993, p. 8). However, Kumar contends that true empowerment can only occur when someone in power loses part of their power. If this is true, what is the impact? Often, community-based projects involve a range of stakeholders where some are poor, marginalized or disenfranchized in some way. The involvement of such groups to create positive change is most likely to result in these groups gaining greater access to services and/or resources, and as a result, to become more empowered. Is a redistribution of power required 'that enables the have-not citizens, presently excluded from the political and economic process, to be included in the future' (Arnstein 1969, p. 216)? The question raised is: do those who were more powerful become less powerful? Or can change foster new ways of working or living together that acknowledge different forms of contribution, and in association, power structures and relationships?

By the late 1990s, the notion that participation and empowerment would lead to faster and more effective progress and development of the community started to take hold, and two ways of engagement evolved. The first relation usually involves a link to the community via one tier, such as a service provider or NGO, which may often be removed, as government is, from the way the community members operate within their lifestyle. Therefore, this tier may either leverage the potential for empowerment or limit it by acting as a filter. Change or development instigated by government agency has often been described as top-down, in that the power brokers set the engagement agenda. A variation of this form of engagement emphasized that market deregulation and institutional reforms deliver development outcomes due to the existence of a civil society. However, this typically has not involved high levels of consultation with the members of that society. When there are two non-government players similar issues arise. Even so, non-government organizations provide a convenient alternative to deliver outcomes.

In contrast, the second way of engagement emphasized 'mobilisation of marginalized groups against the disempowering activities of both the state and the market' (Servaes 2003, p. 248) That is, the community members actively become involved to ensure their lifestyle—for example, practices and cultural beliefs—is not lost and they are still empowered within their community context.

Tandon describes the historical role of citizenship, wherein there arises an obligation to participate in the community, to exercise citizen rights and to engage. This perceived role of the good citizen intersects with the two participation models. These two models of engagement highlight how engagement and participation depend on the principles of citizenship and democracy (Tandon 2008). Tandon suggests that models of participatory programs which are integrated into the democratic political system can be extremely successful in 'enhancing representative forms of democracy' (Tandon 2008, p. 293). However, hegemonies exist within communities, so that the dominant societal paradigm operates even though not all community members are comfortable with the situation. Imbalance between community members in certain societies in regard to resources and access may be understood as the norm. This position does not necessarily flag a need for change for community members. People may accept their lot, while others with access believe the inequities to be fair and reasonable in this particular context. In these instances, non-participation may be seen as appropriate, rather than as demonstrating a lack of citizen involvement.

With the emergence of power-sharing concepts of collaboration, mediation and empowerment (Healey 1992; Wilcox 1994), more radical approaches to community engagement were instigated during the last century. Building on such understandings, in this book we introduce and critique some of the community engagement models, methods and methodologies. We commence our discussion by identifying others' conceptions of community engagement; and thereby, we identify potential implications for those who are considering working in this area.

Frameworks for community engagement and the facilitation of community capacity building have emerged. Kimmel et al.'s (2012) framework highlights how Flora and Flora, in 1993, identified three aspects of 'strong sustainable environments': 'robust physical infrastructure including roads, schools; and human capital such as leaders or access to education; and a strong social infrastructure to facilitate the process of community building and development such as social capital, networks…' (as cited in Kimmel et al. 2012, p. 225).

Kimmel et al. define social infrastructure as the complex and dynamic interactions among people, and through such linkages, 'infrastructure is developed through which information and knowledge are perpetuated within the community, resources and actions are mobilized, and problems are addressed' (2012, p. 226). Strong community social infrastructure depends on the capacity to:

- engage in constructive controversy and devise workable solutions that balance people, place, and economy, rather than divide, exclude, or privilege one portion of a community over others;
- mobilize local capital to invest in regional entrepreneurial activities that benefit a larger community;
- attract and disseminate resources, particularly information, into and throughout a community (Flora and Flora 1993, as cited in Kimmel et al. 2012, p. 226).

Kimmel's group developed a community engagement framework built upon three aspects—holding environments, learning action networks, and social and ecological entrepreneurship. Each is defined as follows:

- Holding environments:

> are safe spaces in which individuals may share ideas freely, adaptive learning can occur, and which frequently rely on facilitation by external partners. A key function of these environments is to direct attention to the problems participants share and to collaborative opportunities to address them. (Heifetz 1994, as cited in Kimmel et al. 2012, p. 227)

- Learning action networks 'are comprised of relationships and linkages among what might otherwise be disparate stakeholder groups and individuals, enabling opportunities for collaboration, mechanisms for exchange and learning, and pathways for shared understanding and mutual benefit' (Stephenson 2011, as cited in Kimmel et al. 2012, p. 227).
- Social and ecological entrepreneurship builds community capacity by supporting entrepreneurship that complements community-identified goals (Kimmel et al. 2012, pp. 230–231).

The shift in power sharing to a model that is more inclusive and to recognize difference within a more equitable relationship is a challenge. That said, the role of user groups or community members is emphasized by Nimegeer et al. (2011) in their discussion of co-creation or co-production, where both service providers and users are recognized as being equal contributors to a situation—a concept originally defined by Alford (as cited in Needham 2008, p. 221). Methods have been employed to foster mutual contribution, but it is recognized that 'while many, including citizen juries and round-table workshops, had merit, it was felt they did not adequately account for the complex value laden and holistic nature of planning, as well as … context' (Nimegeer et al. 2011, p. 1005).

Therefore, in Nimegeer et al.'s Scottish community health study they developed a new planning simulation game: the Remote Service Futures Game. The simulation game uses a number of types and levels of cards, and allows community members, as part of a process of engagement, to express their priorities and designs in a form that is directly usable by health service managers, to contribute to a more applicable service provision. However, although researchers state it as self-evident that co-production of this kind will improve services (Needham 2008), this may not always be the case. To demonstrate, Needham uses social housing to illustrate that co-production requires active engagement, and 'the need for collective dialogue and deliberation between co-producers, rather than purely transactional forms of co-production' (Needham 2008, p. 222).

The process is more than consultation, and Needham's observation reinforces the distinctions made by Christopher Day in 2003 regarding consultation, collaboration and consensus design (Day 2003). '[E]ngaging people in co-production does not happen through consultations, on citizens' juries or at council meetings: it needs to happen at the point of delivery and through conversation and dialogue rather than chance alone' (Parker and Heapy 2006, p. 15). Needham argues that 'the forms of co-production most likely to access therapeutic and diagnostic benefits are those that are collective, dialogical, positive-sum and focused at the point of delivery, rather than individualized, zero-sum and abstracted from service experiences' (Needham 2008, p. 225).

Of interest to those in the field of creating the built environment is the balance between what we see as good design and what may emerge from community consultative and collaborative processes, where the participants have equal and valued input. Fundamental to navigating this tension is a re-evaluation of what 'good' actually means in this context. As Heylighen & Bianchin state, it requires us to drop the preconception that only professionals can identify and appreciate good design (2013, p. 94), and to recognize that the attributes of good design such as 'commodity, firmness, and delight'—defined as far back as the first century BC by the Roman architect Vitruvius (1960)—can be appreciated by all people in relation to their everyday environments and possessions. The way we portray design will also need to be challenged. As Cuff discussed nearly four decades ago (1989, as cited in Heylighen and Bianchin (2013, p. 93), design and architectural journals and magazines reinforce mainstream and entrenched beliefs. This visual emphasis is currently being re-evaluated by spatial designers such as Juhani Pallasmaa (2005). However, the debate also needs to cross over into the community participatory domain, so that high quality design is achieved that is relevant for the community.

Implicated in such discussions are two concepts: community and community capacity. In brief, the former relates to those who participate in a society, while simultaneously constructing what constitutes a community—people, place, beliefs, values, practices, and rituals. Capacity implies the capabilities of both the collective and the individuals in relation to the practices that they desire and need to operate now and in the future. There are many ways these concepts are defined below, and these will be reflected upon to build an understanding of both concepts—community and community capacity—in the context of community engagement and participation.

Social constructs such as community and modes of operating influence how we perceive our places and other people, whether as individuals or as a sub-cultural group. For example, people implicitly enact their lives influenced by the assumptions that they hold, and these beliefs are in part socially constructed. Language also reinforces such understandings:

> That is, the concept of community has evolved from including the usual connotations of coherence and permanence to including the social processes and discourses that create, withstand and dissolve them; in a radio program on Belfast, a local resident, attempting to

> develop non-sectarian initiatives, made a comment along the lines of: 'when you hear the word "community" in this city, you know someone is going to get hurt.' (McManus 2001, p. 45)

Research towards a definition of community has noted that communities must contain 'sufficient social interaction, structure and permanence to allow an individual to identify as a member of that community' (Ragin et al. 2008, p. 1380). This yields a '...sense of membership; common symbol systems; common needs and a commitment to meeting them; and a shared history'; in other words, '...individuals [are] delineated by physical, social, or jurisdictional boundaries' (DiClemente et al. 2002, p. 197).

Thus, communities exist or can come into being because of common physical characteristics such as geographical and spatial boundaries, or by non-physical connections such as shared history or common needs, referred to as an ecological relationship between individual, family and environment (McMurray and Clendon 2011); or due to unforeseen interventions. Thus, physical or emotional bonds become key determinants for the definition of communities. The need to belong to an identifiable group raises the issue of 'others'.

A community may gain definition due to the perceived loss of social coherence of a sub-group of the community in relation to mainstream or dominant community groups; that is, its sense of who it is and how it defines its attributes. For example, rural communities affected by drought leading to high male suicides and rates of depression contradict stereotypical images of the strong country community. Emerging awareness of a sub-group's needs may be as a consequence of increasing social activism, cultural diversity, homelessness, displaced people and/or the need for sustainability (Lommerse 2011, p. 29). In association, 'The most serious political consequence of the desire for community... is that it often operates to exclude or oppress those experienced as different' (Young 1990, p. 234).

Therefore, it is important not to romanticize the concept of community. Communities will operate according to their own rules, and therefore, engaging with them requires knowledge of what these are within the community with which one is working. It may be necessary to ask what is the advantage or disadvantage of a hierarchical community over an inclusive, horizontally structured one for the people involved? And vice versa? Are there circumstances when the existence of one form over the other is not appropriate? For example, the suppression of women, segregation based on skin colour, access to food based on earning capacity, times of war, times of natural disaster, and the like?

In its strongest form, the community can take on structure that further contributes to the sense of community. This structure allows for individuals to partake in decision-making on behalf of their community, in which they are represented and feel ownership (Ragin et al. 2008). The more this structure is developed, the more likely that the community as a whole, as well as the individuals, will endure (Ragin et al. 2008), and therefore, grow collectively. 'Healthy communities are those places where belonging is valued, where the connections between individuals, families and the environments of their lives are as important as the life forces within' (McMurray and Clendon 2011, p. 5).

Because of the fine distinctions within the discourse of communities, community engagement, and participation and capacity building, the chapters within this book are collated into three sections. The first section presents a cross section of projects, which highlight how researchers and practitioners have undertaken community engagement projects across numerous locations and of varying scales and with differing objectives. From this collection it is possible to identify common issues but also insights into how different contexts demand alternative approaches. These involve different methods and can be seen as models for future projects. The second section seeks to reflect upon real life experiences in the field and to offer suggestions and tools for how one may commence projects, and the aspects to address prior to beginning and during the project development. The importance of learning from doing is highlighted. The third section brings together the issues raised in the introduction in light of the examples discussed and the literature. The editors also reflect on future needs and possibilities as a way of moving the debate forward to inform future community-based practices.

Community engagement is conceptualized as 'ways of working'—shaping and managing the environment through the development of strategies, processes, design and construction (Beeck et al. 2011, p. 17). Community engagement is about learning and exchanging knowledge, identifying priorities and possibilities, making decisions, and making things happen (Beeck et al. 2011).

In addition, for us, community engagement is a 'cooperative process of working with people to address their wellbeing, crossing disciplinary boundaries, and using multiple knowledge from inside and outside the community' (Lommerse 2011, p. 26), and as Bryan Bell argues, 'the process of creating the built environment can allow communities and individuals to improve and celebrate their lives. It can help solve their struggles by reshaping their existence and building capacity' (Bell 2008, as cited in Lommerse 2011, p. 26). In this way we aspire to develop and/or describe processes that will facilitate such outcomes; that is, those that are pertinent, meaningful, and positive for all members of a society or community wherever possible.

Capacity Building

Although not evident in all cases, it is assumed that those actively engaging in the majority of community-focused projects are aiming to facilitate positive change for the community members in some way. As part of this agenda-building, the capacity of the people and their community is a central focus. What is meant by capacity? And can it be built in the context of others? An overview of how the concept is defined and has developed follows, in order to establish what the overall goal may look like in relation to these two aspects.

Firstly, capacity building is described as helping people to help themselves (Eade 2007). This may involve incorporating ideas from outside the community. However, the increased or realized capacity emerges from the local context—perspectives, knowledge and skills. Gegeo and Watson-Gegeo (2001) illustrate the concept of empowerment or emancipation in local culture with the story of how this was achieved in Kwara'ae culture, through the integration of indigenous knowledge towards development in the Solomon Islands:

> The Kwara'ae Genealogy Project [is] a research effort by rural villagers aimed at creating an indigenous written account of Kwara'ae culture. In recording, (re)constructing, and writing Kwara'ae culture, project members are not only doing indigenous epistemology, but also reflecting on and critiquing their own indigenous strategies for knowledge creation. (Gegeo and Watson-Gegeo 2001, p. 55)

In this project, community members are able to engage more fully, as their qualities, skills and knowledge enable them to participate differently (Gegeo 1998). This involvement adds to an individual's social capital. Thus, community capacity is not unlike social capital (McMurray and Clendon 2011). Social capital, a concept we expand on below in terms of Pierre Bourdieu's seminal work (1986), is the development of civic engagement, trust and reciprocity amongst community members; the greater the social capital, the more potential there is for improving community capacity (assuming that there are adequate opportunities to leverage from this for the breadth of members of the community).

This increase helps community members to cope with adversity and limitations, as well as fostering a sense of place (McMurray and Clendon 2011). DiClemente et al. (2009) discuss community capacity building in the area of health, while Patel et al. (2011) outline methods for facilitating communication between and within cultural groups to assist cross-cultural engagement. Civic engagement allows power sharing as well as enabling community members to mobilize, thereby collectively improving their quality of life either as individuals or as a collective (Smith et al. 2003).

Secondly, capacity building leads towards transformation—physical, social, economic or environmental—which strengthens our civil society organizations, and can build democratization and strong, accountable governance systems. Thus, capacity is built not only for the local community, but also for the resultant civil society, the policymakers, the professionals and other social actors. Responsibilities and risks are shared, mutual accountability emerges, and partnerships are created that contribute to bring lasting change. Community capacity building is possible within communities residing in both developing and developed countries. Different disciplinary perspectives may be applied. Seminal texts, such as those by Pawar (2005, 2009) and Pawar and Cox (2010), provide examples of community building strategies applicable to the developing and developed world, based on cases that focus on marginalized and excluded communities from India and Australia. Sirianni (2009) discusses the 'bottom-up' approach embedded in the US political system. Whilst focused on governance and politics, the text provides an exploration of the relationship between the term 'democracy' and how it is truly represented in the community.

The disciplines discussed in capacity building literature include social sciences, communication, political studies and development studies. However, a minority have gone beyond their disciplinary boundaries to dialogue with others and to investigate the mutual application and development of their disciplinary community engagement models and methods. Within this book, the integration of multiple disciplines working together in relation to community issues and with community members is of key importance. These range from international contexts, such as rural India, to urban Australian settings. To set the framework for these plural voices of change, we expand on how transdisciplinary methodology can work transformatively for greater community engagement and empowerment.

Reconceptualizing Community as a Discipline in a Transdisciplinary Context

In the context of community engagement projects involving professionals from various discipline bases, community is positioned by the authors as a discipline in its own right; that is, a community has a particular way of operating, just as a discipline does. Professional disciplines and sub-disciplines also mimic communities in that neither grouping is a static entity or skill set, but rather a set of experiences and evolving ideas, as much as they are a set of practices and knowledge.

When disciplines collaborate, the interaction may be described as an inter-disciplinary, multi-disciplinary, cross-disciplinary and/or transdisciplinary activity. If defined as having a distinct set of knowledge, theory, and practices, collaborations involve relating or integrating knowledge and methods from each domain. However, Squire points out that a discipline's culture (philosophy, values, or understandings of the world) is not included in such accounts (Squire 1992, p. 7). In contrast, when disciplines are described as a loose field consisting of many subfields, there may be more opportunities, because subfields within one field may link to core subfields in another (Becher 1990).

Newell's work from the nineteen nineties highlights that if staff of an organization conceptualize inter-disciplinarity in operational terms, they seek to identify discipline topics to cover, references to contribute and the like, when undertaking collaborative projects. In contrast, those having a broader definition seek to understand philosophical differences and reconcile these viewpoints through interdisciplinary analysis and awareness (Newell 1992, as cited in Demirbilek et al. 2007).

Smith et al. (2005) and Demirbilek et al. (2007) provide transdisciplinary studies involving the integration and extension of personal and educational experiences of staff, students, and other stakeholders coming together to design an underground environmental-education centre in Western Queensland. The project contrasted inter-disciplinary activity, which:

> Critically draws on two or more disciplines and…leads to the integration of discipline insights, [with] multi-disciplinary activities, [which] ignore integration, as disciplines make separate contributions… to a problem or issue without an attempt to synthesise…. Cross-disciplinary activity involves two or more disciplines; however, one's principles are applied to the other, or as Newell and Green state, 'one usually exercises the hegemony.' (Smith et al. 2005, p. 7)

In contrast, transdisciplinary activity takes 'as an article of faith the underlying unity of all knowledge' (Newell and Green 1982, p. 24), and thus works beyond the discipline boundaries. Even in the eighties, it was seen that 'the new types of content… produced in response to contemporary world problems, transcend traditional disciplines because of their highly interwoven character and pose methodological problems for curriculum designers and teachers' (Vaideanu 1987, p. 490). 'The current resurgence of the discourse and associated projects involving multiple disciplines would therefore appear to be just as relevant in our contemporary world, which is complex and dynamic' (Smith et al. 2005, p. 8).

The potential to address ill-defined problems through a transdisciplinary approach emerges from the blurred edges of discipline discourses and practices beyond and between the discipline boundaries (Newell 1992). This is the point where new knowledge arises (Peirce Edition Project 1998; Kevelson 1987). This transdisciplinary exchange is similar to the 'co-construction of knowledge', where 'knowledge generation resembles more a collective learning process resulting in a "negotiated knowledge"' (Brand and Gaffikin 2007, p. 287). This 'emancipatory knowledge transcends the blinders created by our conditions and institutions' (Innes and Booher 1999, p. 418).

We propose that the individual community members, alone or as a group, bring their own core knowledge and skills to any project. They are able to sit comfortably within a transdisciplinary context as equal yet unique players and contributors—not just participants or recipients. The bringing of discipline knowledge to a problem in the role of 'expert' can be limiting, and may reinforce current or known approaches and resolutions that restrict processes of consultation. The result may be a compromise rather than reaching the potential of innovative collaborations (Smith et al. 2005). In a holistic situation, professional and personal experiences are combined with common sense understandings (Demirbilek et al. 2007).

This book attempts to answer the following questions:

- How do researchers and professionals engage with the community? What impact on the form and degree of engagement do different disciplinary perspectives, the range of attributes, various scales, and/or different locations and whether it is a developing or developed world context, have?
- Does this community engagement assist to build the capacity of the local community members and empower them socially or economically?
- If so, what are the MODELS and METHODS required for successful or meaningful engagement? What tools are integrated to facilitate the necessary interactions with community?

Models refer to the practices and organization of the way individual researchers and participants have conceptualized, theorized and structured their modes of practice. Embedded within each of these models are a number of methods; that is, the means by which activities are carried out. Collectively they form a suite of tools, which other researchers or project leaders or financiers may draw upon in order to launch or adjust their community engagement projects.

*M*² *Models and Methodologies for Community Engagement* differs from and/or extends others' modes of conceptualization and engagement in this field of practice and research, in that it focuses particularly on the emergence of transdisciplinary approaches; and in doing so, we reposition how the community is conceptualized. To examine this concept more closely, the ideas of Bourdieu are called upon to illuminate our understanding.

Communities within this discussion will be explored as a particular kind of collective or an alternative discipline. By naming it as such we can facilitate opportunities to consider the collective as an entity that is part of the transdisciplinary strategy. We have chosen to conceptualize the community as a discipline in order to stress the equity of the players in a participatory project that stresses community engagement. It is important to premise this discussion with the recognition of a number of attributes of transdisciplinary community action for change:

- the community owns and/or belongs to the location of the project;
- the project activities will occur on community members' territory;
- the project will disrupt the community's existence in some way;
- the consequences (positive and negative) will remain with the community after the life of the project.

As a consequence, the community, as a member of the project, has a unique relationship with it.

That said, the project brings the participants together around a particular set of considerations and associated activities. Therefore, how we conceptualize the relationships involved will influence the tone of the interactions. For instance, projects where external experts arrive to help out will differ according to whether the external experts come to fix a problem defined by parties other than locals, or whether the community brings their expertise and works with external experts to understand and address locally defined issues. In the first case, the local community would be conceptualized (implicitly and/or explicitly) as participants with a deficiency, as participants who are helpless or victimized; in the second case participants would be equal players, or participants who are the hosts and meaningful contributors. To avoid the trap of top-down intervention, we are exploring the concept of a community as a discipline—a collective of people with unique and embedded practices and understandings that they bring to the project as it is explored, understood, and actioned to bring positive change. It should also be noted that disciplines are not static, but fluid, responding to change while also being steeped in history with sub-cultural beliefs and practices. This recognition by the other discipline members of a community's depth of tradition and fluidity accepts the uniqueness of the community and its ability to fully contribute in unforseen ways. As the transdisciplinary agenda looks for new understandings by bringing the unlike together and interrogating the edges between the two, further opportunities arise when multiple players, including the community, are involved. To develop this idea theoretically, we draw on the work of the sociologist Bourdieu on social fields.

Bourdieu identified how collectives of people who share a common understanding operate through a collection of values and beliefs that underpin their actions or practices. For instance, the authors of *M² Models and Methodologies for Community Engagement* belong to fields that include design, health, interior architecture, architecture, social science and planning, in the academy and in practice. Belonging to one or many of these is possible because we hold common understandings, even though the particulars of the field in question may be evolving and the edges blurring with time and experience as well as contextual changes.

Building on Bourdieu's social field theory (Bourdieu 1986), we have conceptualized the community—particularly during projects involving community engagement and participation—as a social practice, and therefore, equivalent to a discipline in the way that it operates. Nolan summarises some of Bourdieu's key concepts as follows.

> …everyday decisions are made within a network of structures and relations, referred to as a field. These decisions shape, and are shaped by, a set of dispositions (habitus) that include attitudes, beliefs, perceptions, and practices, all formed through the embodiment of one's life history. (Nolan 2012, p. 204)

When working with communities, it should be noted that each individual community member's habitus emerges through their life experiences within a social context, each with a particular set of structures and embedded rules. That is, 'habitus is a whole body experience, but it is not one that can be attributed solely to an individual since dispositions are created and recreated through social interaction and tradition' (Nolan 2012, p. 204); moreover, ongoing experience of both the game that constitutes how we go about everyday life, and the structures in which it occurs, enables orientation and redirection (Nolan 2012). The embedded power relations in a society are linked to the existence of cultural and economic capital. Bourdieu describes cultural capital as the means for individuals to gain, have and maintain recognition, thereby positioning themselves strategically within the field (1986). And in turn, this position gives access to economic capital, which gives power (Bourdieu 1986; Nolan 2012).

A community, if understood as a field, needs to be considered within the context of capacity building being a transformative experience that can occur within a given field. Bourdieu uses the term *doxa* for that which is taken for granted in a society—'the pre-verbal taking-for-granted of the world that flows from practical sense' (Bourdieu 1990a, p. 120); he emphasizes that the core values and discourses of a social practice field are not evident or questioned even if they may not always be logical (Nolan 2012, p. 205). Where the individuals are positioned in the field will provide differing opportunities to engage, to be seen to be engaged, and to influence the modes of engagement. The modes of operation in these positions are most often implicit and a way of being, yet can bring transformation. There is 'commitment to the presuppositions—*doxa*—of the game' (Bourdieu 1990b, p. 66).

This observation can be applied to the community but also to the discipline; the person—the planner, for example—who is not a local but is engaging in the process is inherently an outsider to the community, but operates tacitly within his or her own field. As Bourdieu could be imagined to have stated, they do not necessarily know the game of the others, nor how to 'compete' within it. Nolan summarizes:

> …the relative smoothness of the game/field often depends upon the players blindly accepting and following these rules, regardless of how arbitrary they might seem. As one continues to engage in the game, the rules seem natural and unquestionable to the players, resulting in a "feel for the game" which no longer requires the deliberate act of thinking carefully about each and every move before acting. However, should players not follow the rules as laid out before them, penalties will surely follow. (Nolan 2012, p. 204)

To envisage change in such a context and to predict transformation requires sensitivity and careful negotiation, including reflection upon the nature of the discipline groups as fields, and their habitus, in addition to the community.

Previously in this chapter, social capital was introduced in relation to capacity building as a potential aim, or at least, potential outcome. Social capital needs to be considered as part of the way a community operates as a field, and therefore, is integrally linked to structures, resources, rules, discourses and values. As Wakefield and Poland state:

> …counter to many current policy discourses, which focus on the importance of connection and cohesion without addressing fundamental inequities in access to resources… approaches to community development and social capital should emphasise the importance of a conscious concern with social justice. (2005, p. 2819)

Therefore, as partners in community projects, we need to be cognizant of the underlying balances and tensions, which may not be immediately evident, but that may influence 'the game'.

Wakefield and Poland (2005) discuss three themes or aspects of social capital that are relevant for those in this area of work, building on those originally described by Bourdieu (1986). Firstly, the communitarian aspect, drawing on Putnam's definitions of social capital contextualized in relation to neighbours, clubs, and community as:

> the networks, norms, and social trust that facilitate co-operation for mutual benefit…. [Putnam] asserts that societies with high levels of interpersonal trust, prosocial norms, and interpersonal networks that emphasize reciprocity are more likely to experience positive economic, political, and social development than those lacking these characteristics. (Wakefield and Poland 2005, p. 2821)

In this ethos, societies benefit at a more personal level, as people are more likely to assist one another without a need for personal reward, and social cohesion is facilitated. In terms of capacity building, the implicit networks and cohesion may be used constructively, and/or they may negate relationships with outsiders. Therefore, we posit that understanding how and why a community has internal trustworthy networks, and what interventions may cause their disruption, is a worthwhile consideration and mission.

Secondly, the institutional aspect of social capital emphasizes 'not social ties per se, but rather the scaling up of these ties to form organizations that are politically and economically effective' (Wakefield and Poland 2005, p. 2821). The emphasis is the link between the local acts and people with the larger organizational operations; and that link in turn can lead to broader benefits for more people (Wakefield and Poland 2005). In Wakefield and Poland's description of Rothstein's research into Swedish democracy, this link is clear; the interest group organizations (that is, local scale) worked with the State to form policies (institutional scale), which when implemented, have led to social welfare and civic involvement, from which the community (broad scale) has benefited and developed. This emphasizes the need (or the possibility) to leverage from the immediate concerns of participants into the broader context of a project in order to create greater impact. However, Wakefield and Poland also point out that the formation of ties into effective organizations needs synergies and meaningful co-operations between the community and the institutions, rather than top-down directives or forced control.

The third aspect of social capital, power relations, is the critical one. As Wakefield & Poland state (building on Boudieu's work), 'resources linked to the possession of durable networks of acquaintance and recognition… emerge from, and [are] structured by, other forms of capital, particularly economic' (Wakefield and Poland 2005, p. 2822).

> …an apparently gratuitous expenditure of time, attention, care, concern, which, as is seen in the endeavor to personalize a gift, has the effect of transfiguring the purely monetary import of the exchange and, by the same token, the very meaning of the exchange. From a narrowly economic standpoint, this effort is bound to be seen as pure wastage, but in the terms of the logic of social exchanges, it is a solid investment, the profits of which will appear, in the long run, in monetary or other form. Similarly, if the best measure of cultural capital is undoubtedly the amount of time devoted to acquiring it, this is because the transformation of economic capital into cultural capital presupposes an expenditure of time that is made possible by possession of economic capital. (Bourdieu 1997, p. 54)

Of interest here is Bourdieu's observation that our appearance and behaviour can either provide us access, to or restrict us from, particular situations or subcultural groupings:

> Exchange transforms the things exchanged into signs of recognition and, through the mutual recognition and the recognition of group membership which it implies, reproduces the group. By the same token, it reaffirms the limits of the group, i.e., the limits beyond which the constitutive exchange trade, commensality, or marriage cannot take place. Each member of the group is thus instituted as a custodian of the limits of the group: because the definition of the criteria of entry is at stake in each new entry, he can modify the group by modifying the limits of legitimate exchange through some form of misalliance. (Bourdieu 1997, p. 52)

Therefore we need to personally invest effort, energy and time to stay in-group, or to move between groups, depending on the social setting. Within societies with greater capital there are more opportunities to change. For example, as in the case of one of the authors, the Australian rural farm child can go to university to become a scientist, to retrain as an architect and then reposition that accumulated knowledge and skills in an academic career. The freedom to move between fields and to develop a habitus embedded in the experience has led to the acquisition of transferable and specific forms of social capital. However, for a female in a rural Indian village, would these opportunities be possible? How can community engagement projects facilitate such acquisitions? And what are the consequences if they do?

Wakefield and Poland highlight how social processes can engender:

> '... identification, normalization and marginalization', and that 'groups in lower social positions are most likely to experience the branding of their activities as base, uncouth, ignorant, or worse, pathological and deviant' (Bourdieu 1979; Bauder 2002), while those in higher social positions experience their activities as socially valued. This process is contingent and contested—classificatory systems are sites of struggle between groups they characterize, and social groups strive to both change existing characterizations and to turn them to their advantage. (Bourdieu 1979; Mansbridge 2001; Wakefield and Poland 2005, p. 2826)

When we apply these insights to community projects, questions are raised regarding what processes, generated through the project, expose or conceal the embedded relationships. If capacity building is the aim, then how do the implicit structures and positions become disturbed, and what are the likely consequences? In the past, stories of the researchers as saviours in 'under'-developed countries demonstrated a lack of sensitivity to the nature and worth of the rituals and practices, as well as to the appearance and language of the communities that were studied (Choguill 1996). Those working alongside, for, or with communities still have the potential to repeat these past habits if the aspects of social capital are not reflected upon. Similar awareness is needed when addressing the concepts and practices of transformation and building capacity.

Exploring the Issues in Practice: the Chapters

Each of the chapters within this book demonstrates an approach to working with community. Each has distinct intentions and practices. The collection as a whole reflects on how community has been positioned, and whether the projects studied have been transformative and/or have facilitated positive capacity development. The following diagram captures the scope of the field as we conceived it to be when developing this book. Each project will navigate the field and overlay it in a distinct way. Collectively, the values, beliefs, attitudes, resources and practices will provide insights into this particular field (Fig. 1.1).

The book is divided into three broad sections—disciplinary approaches; cross-disciplinary approaches; and beyond, defining transdisciplinary insights through engagement with people as key players. All are linked to the realm of community capacity building.

Part I includes reflective narratives and commentary on different disciplinary approaches to community capacity building. Capacity building within the context of this book refers to empowering the community members involved, the researchers facilitating the project and/or the organizations to which they belong. The nine chapters that follow present perspectives from practitioners and researchers in the fields of health, social sciences, fashion, design, anthropology, natural resources, interior architecture, architecture, urban and regional design and planning. Their case studies provide a varied, international overview set in urban, rural and remote places in India, Bangladesh, Australia, Vanuatu, Uganda, Japan, Sri Lanka, Spain

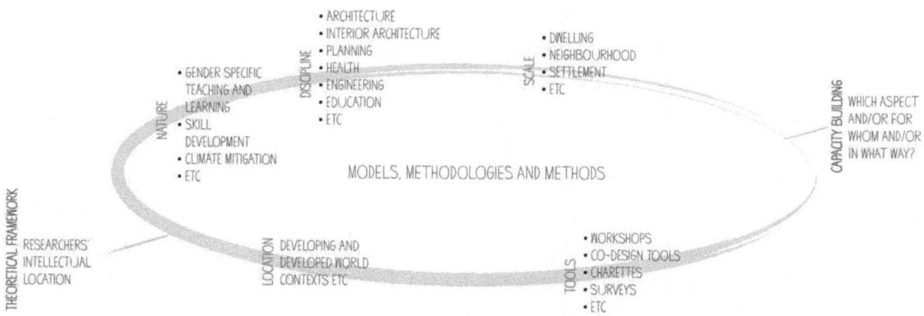

Fig. 1.1 Community engagement framework: models, methodologies and methods

and the United States. The case studies vary from micro- to macro-scale community engaged projects.

Chapters 2–4 focus on how community engagement projects may occur. The second chapter is about a local community in India using local resources, knowledge and technologies for the betterment of the community as a whole. The third chapter presents the importance of decision-making for capacity building through a community study with women and men from rural Uganda. The fourth chapter explores the premise that citizen participation should reinforce community empowerment in neighbourhood regeneration, through the analysis of two significant neighbourhood renewal experiences developed in the Barcelona Metropolitan Region. Featured in the fifth chapter is a case study in Vanuatu, using a design/build model to facilitate income generation for the local community, resulting in capacity building for Vanuatu community members and architectural students.

Chapter 6 explores communities of practice, using a case study in India to illustrate how Participatory Action Research promotes mutually respectful relationships, shared responsibilities, and an emphasis on local capacity building, thereby promoting environments in which communities increase their ability to uncover local barriers and harness local assets to build healthier communities. Chapters 7–10 explore communities of knowledge and learning. Chapter 7 outlines the new attributes of leadership needed to address projects based in the contemporary world, and draws on nature as an inspiration.

The eighth chapter describes a small intervention in suburban Australia that used pro-bono work to facilitate authentic graduate learning, capacity-building opportunities and community engagement. The ninth chapter, again set in urban Australia, describes a community design project which explored the potential for work-integrated learning experiences to enliven communities in a major exhibition-cum-festival.

Part I finishes with the tenth chapter, which explores a case study in Japan where the artist/designer used inclusive approaches to community-based projects to ensure long-term success and navigate risk.

Part II is an exploration of the common ground between the different disciplinary approaches. Can the above disciplinary models and embedded methods offer insights into transdisciplinary opportunities with wider application? The chapters in Part II strive to capture any interrelationships that become obvious through the juxtaposition of the case studies related in Part I. Chapter 11 visually diagrams the models embedded in the case studies and calls for the reader to identify connections that may be meaningful for him or her. The disciplinary and locational variables are also mapped against the framework presented in this introductory chapter to demonstrate the potential threads present in the case studies, and thereby opportunities embedded in the collective example for new insights.

The next two chapters offer lessons learnt from reflections upon a number of project scenarios. Chapter 12 provides examples of the barriers that may arise during community projects ranging in scale from micro interior projects to larger village and town settings. Chapter 13 relates experiences from several action research and academic studies of natural resource management in Bangladesh.

Chapter 14 presents a new tool, formulated as an instrument to identify and categorize context-specific challenges. It is intended to help stakeholders identify potential road blocks in their project, which will give them options to develop proactive strategies.

Part III reviews: how researchers understand and investigate the idea of community; whether and how the definition of ‘community’ by the researchers is dependent on location and nature of projects; how researchers’ views change through reflection and interaction with community members; and what constitutes capacity building in different socio-cultural contexts. The case studies described in the chapters in Parts One and Two are reflected on in terms of the concept of community as a discipline in its own right, which builds capacity in a transformative experience that brings physical, social, economic or environmental change. We revisit notions of democracy and development/transformation, and identify transdisciplinary understandings of community engagement.

Finally, we review the process by which this book was developed, as a model in itself of capacity building.

As we stated in the introduction, multiple possibilities are presented in this work, some challenging and others rewarding. The projects discussed capture a multitude of ways to interface and interact with communities, of modes of interaction, empowerment or engagement, of varying scales and complexity and socio-economic circumstances, as well as of physical locations across the world. We aimed to consider what it is about community involvement that not only attracts some professionals into working this way, but also what makes some community work rewarding for the professional, and—more importantly—successful from the community members’ perspective. The case studies and narratives that you are about to read will answer these questions, while the chapter authors demonstrate their passion for working with communities to achieve positive outcomes in their very varied forms.

Acknowledgements Many thanks to Jessica Winters for contributing to the literature review.

References

Arnstein, S. (1969). A ladder of citizen participation. *Journal of the American Institute of Planners, 35*(4), 216–224.

Bauder, H. (2002) Neighbourhood effects and cultural exclusion. *Urban Studies, 39* (2002), 85–93.

Becher, T. (1990). The counter-culture of specialization. *European Journal of Education, 25*(3), 333–346.

Beeck, S., Smith, D., Lommerse, M., & Metcalfe, P. (2011). An introduction to social sustainability and interior architecture. In D. Smith, M. Lommerse, & P. Metcalfe (Eds.), *Life from the inside: Perspectives on social sustainability and interior architecture* (pp. 24–57). Perth: Pencil and Paper.

Bell, B., & Wakeford, K., (Eds.). (2008). *Expanding architecture: design as activism,* Metropolis Books, New York: Greenwood.

Bourdieu, P. (1979). *Distinction: a social critique of the judgement of taste.* London: Routledge. Trans. R. Nice.

Bourdieu, P. (1986). The forms of capital. In J. Richardson (Ed.), *Handbook of theory and research for the sociology of education* (pp. 241–258). New York: Greenwood.

Bourdieu, P. (1990a). *In other words: Essays toward a reflexive sociology.* (trans: Loïc J. D. Wacquant) Stanford: Stanford University Press.

Bourdieu, P. (1990b). *The logic of practice* (trans: R. Nice). Cambridge: Polity Press.

Bourdieu, P. (1997). The forms of capital. In A. H. Halsey, H. Lauder, P. Brown, & A. S. Wells (Eds.), *Education: Culture economy and society* (pp. 46–58). Oxford: Oxford University Press.

Brand, R., & Gaffikin, F. (2007). Collaborative planning in an uncollaborative world. *Planning Theory, 6*(3), 282–313.

Choguill, G. A. (1996). A ladder of community participation for underdeveloped countries. *Habitat International, 20*(3), 431–444.

Cuff, D. (1989). The ethos and circumstance of design. *The Journal of Architectural and Planning Research,* 6(4). (Winter, 1989):305e320.

Day, C. (2003). *Consensus design: Socially inclusive process.* Princeton: Architectural Press.

Demirbilek, N., Smith, D., Scott, A., Dawes, L., & Sanders, P. (2007, April). *The role of the learning context and designer characteristics challenge and reveal the design process.* Paper presented at Dancing with disorder: Design, discourse, & disaster, EAD07 Conference, Izmir, Turkey.

DiClemente, R. J., Crosby, R. A., & Kegler, M. C. (Eds.). (2002). *Emerging theories in health promotion practice and research: Strategies for improving public health.* San Francisco: Jossey-Bass.

DiClemente, R. J., Crosby, R. A., & Kegler, M. C. (Eds.). (2009). *Emerging theories in health promotion practice and research* (2nd ed.). San Francisco: Jossey-Bass.

Dudley, E. (1993). *The critical villager: Beyond community participation.* London: Routledge.

Eade, D. (2007). Capacity building: Who builds whose capacity? *Development in Practice, 17*(4/5), 630–639.

Flora, C., & Flora, J. (1993). *Entrepreneurial social infrastructure: A necessary ingredient.* Annals of American Academy of Political and Social Science, 529(1), 48–58.

Gegeo, D. (1998). Indigenous knowledge and empowerment: Rural development examined from within. *The Contemporary Pacific, 10*(2), 289–315.

Gegeo, D., & Watson-Gegeo, K. (2001). "How we know": Kwara'ae rural villagers doing indigenous epistemology. *The Contemporary Pacific, 13*(1), 55–88. doi:10.1353/cp.2001.0004.

Hall, M. (2010). Community engagement in South African higher education. *Kagisana, 6.* Durban, SA: Council on Higher Education. http://ahero.uwc.ac.za/index.php?module=cshe & action=downloadfile & fileid=18409092513316751038744.

Healey, P. (1992). Planning through debate: The communicative turn in planning theory. *Town Planning Review, 63*(2), 143–159.

Heifetz, R. A. (1994). *Leadership without easy answers.* Cambridge, MA: Cambridge University Press.

Heylighen, A., & Bianchin, M. (2013). How does inclusive design relate to good design? Designing as a deliberative enterprise. *Design Studies, 34*(1), 93–110.

Innes, J., & Booher, D. (1999). Consensus building and complex adaptive syatems: A framework for evaluating collaborative planning. *APA Journal, 65*(4), 412–423.

Israel, B. A., Schulz, A. J., Parker, E. A., & Becker, A. B. (1998). Review of community-based research: Assessing partnership approaches to improve public health. *Annual Review of Public Health, 19,* 173–202. doi:10.1146/annurev.publhealth.19.1.173.

Israel, B. A., Schulz, A. J., & Parker, E. A. (2001). Community-based participatory research: Policy recommendations for promoting a partnership approach in health research. *Education for Health, 14*(2), 182–197. http://ejournal.narotama.ac.id/files/Community-Based Participatory ResearchPolicy.doc. Accessed 30 Aug 2013.

Kevelson, R. (1987). *Charles S. Peirce's method of methods: Vol. 17. Foundation of semiotics.* Amsterdam: John Benjamins.

Kimmel, C. E., Hull, R. B., Stephenson, M., Robertson, D., & Cowgill, K. (2012). Building community capacity and social infrastructure through landcare: A case study of land grant engagement. *Higher Education, 64*(2), 223–235.

Kumar, S. (2005). Community programmes in palliative care: What have we learned? *Indian Journal of Palliative Care, 11,* 55–57.

Kumar, S., Chen, L. C., Choudhury, M., Ganju, S., Mahajan, V., Sinha, A., et al. (2011). Financing health care for all: Challenges and opportunities. *Lancet, 377*(9766), 668–679.

Lommerse, M. (2011). Working together: Interior architecture creating the community. In D. Smith, M. Lommerse, & P. Metcalfe (Eds.). *Life from the inside: Perspectives on social sustainability and interior architecture* (pp. 24–57). Perth: Pencil and Paper.

Mansbridge, J. (2001). The making of oppositional consciousness. In: J. Mansbridge, A. Morris (Eds.), *Oppositional consciousness*, University of Chicago Press, Chicago (2001).

McManus, P. (2001). One step forward and two steps back: Urban policy and community planning in England since 1979. In O. Yiftachel, J. Little, D. Hedgcock, & I. Alexander (Eds.), *The power of planning: Spaces of control and transformation* (pp. 45–56). Dordrecht: Kluwer Academic Publishers.

McMurray, A., & Clendon, J. (2011). *Community health and wellness: Primary health care in practice* (4th ed.). Chatswood: Elsevier Australia.

Minkler, M., & Wallerstein, N. (Eds.). (2008). *Community-based participatory research for health: From process to outcomes* (2nd ed.). San Francisco: Jossey-Bass.

Mooney, G. (2005). Communitarian claims and community capabilities: Furthering priority setting? *Social Science & Medicine, 60*(2), 247–255.

Needham, C. (2008). Realising the potential of coproduction: Negotiating improvements in public services. *Social Policy and Society, 7*(2), 221–231.

Newell, W. H. (1992). Academic disciplines and undergraduate interdisciplinary education: Lessons from the school of interdisciplinary studies at Miami University, Ohio. *European Journal of Education, 27*(3), 211–221.

Newell, W. H., & Green, W. J. (1982). Defining and teaching interdisciplinary studies. *Improving College and University Teaching, 30*(1), 23–33.

Nimegeer, A., Farmer, J., West, C., & Currie, M. (2011). Addressing the problem of rural community engagement in healthcare service design. *Health & Place, 17*(4), 1004–1006.

Nolan, K. (2012). Everyday decisions. *Educational studies in mathematics, 80*(1), 201–215.

Pallasmaa, J. (2005). *The eyes of the skin: Architecture and the senses* (2nd ed.). Hoboken: Wiley.

Parker, S., & Heapy, J. (2006). *The journey to the interface: How public service design can connect users to reform.* London: Demos. http://www.demos.co.uk/files/journeytotheinterface.pdf. Accessed 30 Aug 2013.

Patel, F., Li, M., & Sooknanan, P. (2011a). *Intercultural communication: Building a global community.* New Delhi: Sage. doi:10.4135/9781446270318.

Patel, V., Kumar, A. K. S., Paul, V. K., Rao, K. D., & Reddy, K. S. (2011). Universal health care in India: The time is right. *Lancet, 377*(9764), 448–449. doi:10.1016/S0140-6736(10)62044-2.
Pawar, M. S. (Ed.). (2005). *Capacity building for participation: Social workers' thoughts and reflections*. Wagga Wagga: Centre for Rural Social Research.
Pawar, M. S. (2009). *Community development in Asia and the Pacific*. New York: Routledge.
Pawar, M. S., & Cox, D. R. (2010). *Social development: Critical themes and perspectives*. New York: Routledge.
Peirce Edition Project. (Ed.). (1998). *The essential Peirce: Selected philosophical writings (Vol. 2, 1893–1913)*. Bloomington: Indiana University Press.
Ragin, D. F., Ricci, E., Rhodes, R., Holohan, J., Smirnoff, M., & Richardson, D. R. (2008). Defining the "community" in community consultation for emergency research: Findings from the community VOICES study. *Social Science & Medicine, 66*(6), 1379–1392.
Servaes, J. (2003). *Approaches to development: Studies on communication for development*. Paris: UNESCO.
Sirianni, C. (2009). *Investing in democracy: Engaging citizens in collaborative governance*. Washington D. C.: The Brooking Institution.
Smith, N., Littlejohns, L. B., & Dimple, R. (2003). *Measuring community capacity: State of the field review and recommendations for future research. Alberta*. Canada: Health Policy Research Program.
Smith, D., Sanders, P., Demirbilek, N., & Scott, A. (2005). *Designing together: A collaborative experiment in design methodology within a multi-disciplinary environment*. Paper presented at Drawing together: Convergent practices in architectural education, 3rd Conference of AASA, Brisbane, Queensland.
Squire, G. (1992). Interdisciplinarity in higher education in the United Kingdom. *European Journal of Education, 27*(3), p.201–10.
Stephenson, M. (2011). Conceiving land grant university community engagement as adaptive leadership. *Higher Education, 61*(1), 95–108.
Tandon, R. (2008). Participation, citizenship and democracy: Reflections on 25 years of PRIA. *Community Development Journal, 43*(3), 284–296.
Tsey, K. (2008). Tackling endemic substance abuse among Indigenous Australians: The contribution of values-based family empowerment education. *Forum on Public Policy* (A Journal of the Oxford Round Table): Urbana.
Tsey, K., Harvey, D., Gibson, T., & Pearson, L. (2009). The role of empowerment in setting a foundation for social and emotional wellbeing. *The Australian e-Journal for the Advancement of Mental Health, 8*(1), 6–15.
Vaideanu, G. (1987). Interdisciplinarity in education: A tentative synthesis. *Prospects, xvii*(4), 487–501.
Vitruvius, M. (1960). *The ten books on architecture*. (trans: M. H. Morgan). New York: Dover.
Wakefield, S., & Poland, B. (2005). Family, friend or foe? Critical reflections on the relevance and role of social capital in health promotion and community development. *Social Science & Medicine, 60*(12), 2819–2832.
Wallerstein, N. B., & Duran, B. (2006). Using community-based participatory research to address health disparities. *Health Promotion Practice, 7*(3), 312–323.
Wallerstein, N. B., & Duran, B. (2010). Community-based participatory research contributions to intervention research: The intersection of science and practice to improve health equity. http://ajph.aphapublications.org/doi/abs/10.2105/AJPH.2009.184036. Accessed 22 June 2013.
Wilcox, D. (1994). *The guide to effective participation*. London: Joseph Rowntree Foundation.
Young, I. M. (1990). *Justice and politics of difference*. Princeton: Princeton University Press.

Part I
Narratives on Models, Methodologies and Methods

This section of *M*^{2} is about sharing and learning. *M*^{2} positions the contributors as a collective; people who have come together as a group through their belief that the majority of community members have the potential to contribute to society and join in the practice of community engagement. The aim of the chapters is to improve the potential and quality of interactions like these in future projects. The diversity of the contributors and their selected projects challenges traditional definitions of community participation, engagement and capacity, and contributes to the construction of a discipline of community practice that suits twenty-first century contexts across the globe. The projects are offered for the reader to critique and to reflect upon.

> I love reading another reader's list of favorites.
> Even when I find I do not share their tastes or predilections,
> I am provoked to compare, contrast, and contradict.
> It is a most healthy exercise, and one altogether fruitful.
> T.S. Eliot, *The Four Quartets*

Chapter 2
An Ethnographic and Collaborative Model of Inquiry: Activity Centre Project in India

Reena Tiwari and Yatin Pandya

Abstract This chapter proposes an ethnographic and collaborative model of inquiry for executing community scale architectural projects that facilitate capacity building opportunities for all stakeholders involved in the process. Taking on the role of an ethnographer, the designer/architect/researcher is able to locate her- or himself within the time and space of the community. In the process he/she is able to form an insider's perspective on issues and challenges. Working collaboratively, the intention is to create a horizontal power relation between different stakeholders. The model is described in general terms, then illustrated by a case-study—an Activity Centre project for an Indian slum settlement—then discussed in terms of its capacity building outcomes.

The Activity Centre project was initiated by a non-government organization, and involved the local architect, the academic researcher and architecture students, residents of a slum settlement, and skilled labour. Although each stakeholder entered into the project with different objectives, the project outcomes were able to provide benefit to all in a tangible or intangible manner. The chapter reflects on the methods employed in the project and the role of stakeholders at each stage of the process. The importance of the local context is underscored—local people, local resources, local tools and techniques—during the design and construction process, to achieve an outcome that is innovative and engenders reciprocity and learning amongst the stakeholders.

Keywords Ethnography · Collaborative planning · Capacity building · Economic empowerment · Gender equity

R. Tiwari (✉)
Departments of Urban and Regional Planning, and of Architecture and Interior Architecture, Curtin University, Perth, Australia
e-mail: r.tiwari@curtin.edu.au

Y. Pandya
FOOTPRINTS Environment Architecture Research Technology Housing (E.A.R.T.H), Ahmedabad, India
e-mail: pandyatin@hotmail.com

R. Tiwari et al. (eds.), *M² Models and Methodologies for Community Engagement,*
DOI 10.1007/978-981-4585-11-8_2, © Springer Science+Business Media Singapore 2014

Introduction

The chapter discusses an ethnographic and collaborative approach to capacity building used in a project in an Indian city (2004–2008). A group of students, led by Reena Tiwari (professor/researcher) from the School of Built Environment, Curtin University Australia, worked in collaboration with non-government organizations and other local stakeholders in India on an 'Activity Centre' project led by Yatin Pandya (local architect at Vastu Shilpa Foundation at that time). The chapter first outlines a model of enquiry that emerges from an ethnographic and collaborative planning context. This is followed by an overview of the methods and methodology used within this model in the project. After reflecting on the capacity building outcomes of the project and assessing its success, the chapter concludes by stressing that researchers and practitioners need to recognize the potential of the locale in which they work—its resources, its people and their knowledge—for creating and strengthening identity and enhancing a sense of community.

Model of Inquiry

In response to the failure of the traditional top-down planning approach, the idea of collaborative planning was introduced and launched by Healey (1997). Her aim was to overcome the hierarchical and bureaucratic framework that attended centralized decision-making. Contemporary approaches to planning have included citizens in the decision-making process (Healey 1999; Tewdr-Jones and Allmendinger 1998) and thus have brought many positive outcomes to projects, for the following reasons:

- Local residents help to bring an insider's perspective to the problems;
- Because local needs are able to be identified through community involvement, the deliverables of the program are worthwhile and acceptable to the community;
- Program outcomes that involve community deliberations have longevity, since there is a sense of community ownership of them;
- Community involvement builds local organizational skills, and the capacity to develop strong successor organizations (Chaskin 2001; Ife 1995).

On the other hand, by resisting community consultations, governing agencies force local residents to respond negatively to proposed programs (Healey 1997). It is sometimes claimed by agencies that consultation becomes pointless, as residents lack the ability to understand strategies; however, as Davoudi and Healey point out, the problem here lies more in the way consultation is envisaged and carried out (1995). To address this, Healey proposes:

> … planning processes need to work in ways which interrelate technical and experiential knowledge and reasoning, which can cope with a rich array of values, penetrating all

> aspects of the activity and which involve active collaboration between experts and officials in governance agencies and all those with a claim for attention arising from the experience of co-existence in shared places. (1997, p. 87)

The conception and discharge of consultation should be a social process built on an understanding of the social values and social relations already existing in the place. Developing a social understanding involves an experiential and a situationist perspective (Debord 1977), where consultants/researchers are required to take an ethnographic stand by immersing themselves in the lives of the people, and need to place the phenomena studied in their social and cultural context (Lewis 1985, p. 380). Ethnography relies on locating the researcher firmly within the time and space of social actors, and thus is able to reveal the socio-scape that people collectively construct (Gille and Riain 1995). As Jon Kolko says, ethnography is about leaving the safety of the studio and going out into 'real life'; it is about observing 'real people' and designing [and working] with them (Kolko 2011, p. 50). The aim 'is to provide rich, holistic insights into people's views and actions, as well as the nature (that is, sights, sounds) of the location they inhabit, through the collection of detailed observations and interviews' (Reeves et al. 2008, p. 512). This allows consultants and researchers to reach an insiders' understanding, or 'get inside' the way each group of people sees the world (Hammersley 1992).

There are different views on the distance that needs to be maintained between the local residents and the outside researcher/consultant architect. If the goal is to study the phenomenon/conditions within the context, traditional notions rest on researchers' detachment rather than engagement, so as not to influence or mediate relations and knowledge existing in the local space (Reeves et al. 2008). However, in projects where the goal is to empower local residents and facilitate social change, or where (as in the Activity Centre project that is subject of this chapter), the goal is to co-design[1] and co-construct physical interventions in the built form, incorporating local socio-physical and economic aspects, the researchers have to play the role of action ethnographers (Nilsson 2000). Here they have to be more than participant observers and become change agents, facilitating local adaptations and change (Fig. 2.1). 'This role [participant observers] positions us [them] outside the context or organization (having "peripheral membership"), while the role of change agent positions us [them] inside the organization (having "active membership")' (Adler and Adler 1997, as cited in Barab et al. 2004, p. 255).

Therefore, our model of enquiry and investigation is collaborative and ethnographic, with a goal of facilitating change and adaptation for all stakeholders (local residents, researchers, NGOs and local professionals) through co-design and co-construction. The involvement of all stakeholders in the design and construction stages of the project enables them to imbibe local knowledge and resources and respond to local needs. The following sections describe the project and the methodology and methods used.

---

[1] Within a participatory design framework, co-design is considered to be 'collective creativity of collaborating designers' or 'the creativity of designers and people not trained in design working together in the design development process' (Sanders and Stappers 2008, p. 6).

Fig. 2.1 Action ethnographer

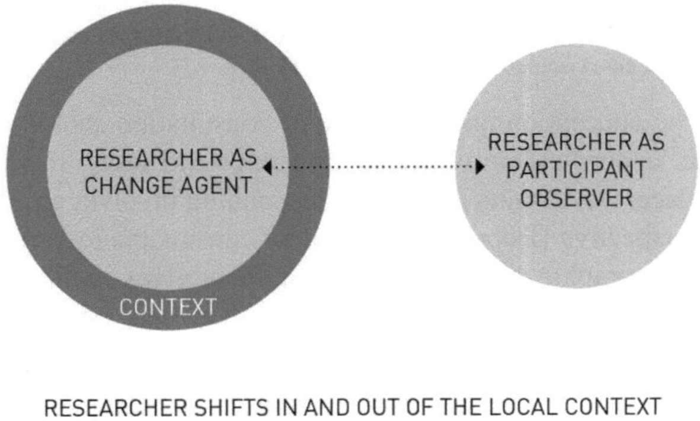

Fig. 2.2 Randev Pir Tekro, Ahmedabad

Project and Methodology

A community action project was carried out from 2004 onwards for a slum community (Ramdev Pir Tekro) in the city of Ahmedabad in India (Fig. 2.2). Tekro is home to more than 150,000 people (Pandya 2008). The slum residents earn their living by driving rickshaws, cleaning homes, shining shoes and working as skilled and unskilled labourers. The majority of residents work as rag-pickers, especially young girls and women, in a waste recycling industry that is the main economic driver for the settlement. The land belongs to the government, and Tekro dwellers have illegally squatted for the past 50 years and more. Initially, there was no provision of basic infrastructure services such as electricity, footpaths, drainage, water supply and so on. The situation has slightly improved with the aid of NGOs and government interventions; however, more is needed in order to resolve continuing and increasing health issues, inadequate toilet and sewage facilities, poor conditions of housing, low incomes, high-interest loans, child labour and illiteracy (Manav Sadhana n.d.).

Manav Sadhana, an NGO with a strong local base and global connections, has been working towards promoting values, education and awareness among children and economic empowerment of women of the Tekro, with the objective of improving the living standards of the dwellers. One of its initiatives has been the Activity

Centre project that started in 2004 and was completed in 2006. The completed project, a multi-purpose activity centre, now serves as a community centre with health camps, festivity ground, gym, and importantly, an informal school for young children. It also provides evening education for adults, and is a vocational training and activity workshop for the manufacture of craft-based products by women and the elderly. These handmade craft-based products include paper bags, greeting cards, diaries, albums, envelopes and gift hampers, which are sold under the label 'Manav Sadhana Manufacturers', and the profit earned is distributed among the children and women workers. This enabling project, initiated by Manav Sadhana for the women, elderly and the children, is known as 'Earn and Learn' and is credited with making a major difference to the lives of the poor. The campus also includes a dormitory, an administrative unit and an all-religion meditation unit (Trivedi and Tiwari 2010).

The above project started in 2004. Curtin University students participated in two stages of the project—needs assessment and concept design—in 2005. For students, this project was also going to be assessed as part of their course. They worked with Yatin Pandya, who led the project to completion in 2006. This project won awards from the World Architecture Community (WAC), the Squat City Urban Inform Forum at Rotterdam (World Architecture Community 2009) and several national organizations.

Project Objectives

Different stakeholders had different objectives for the project. Since a huge waste recycling industry flourishes within the settlement, the main objective of the project as defined by the architect was to develop innovative building components that use waste, simple hand-operated tools and local resources and know-how. For Manav Sadhana, the project objectives had a philanthropic component that aimed to address the issues faced by the community, while for the Curtin researcher and students the objectives were 'learning by doing' and community facilitation through engagement. The community's interests and objectives were represented by Manav Sadhana, as it had been closely working with the community for several years.

Within a collaborative and ethnographic model of inquiry, our methodology incorporated various stages, as shown in Table 2.1. It started with building relationships with the local community. Community needs were assessed through observations and surveys, informal chats and structured questionnaires. This stage was followed by conceptual design work and subsequent discussions on detailed designs. Building components were then tested, since new building prototypes using municipal waste were being designed. Construction started in 2005 and the Centre was formally opened in 2006.

The project's collaborative planning had the goal of creating a horizontal power relation between different stakeholders, and to this end, a wide range of tools was employed, like meetings, surveys, observations, informal chats, co-construction, community events, and awareness programs (Table 2.2).

Table 2.1 Involvement of different stakeholders in different stages of the project

STAGES	METHODS	PARTICIPANTS INVOLVED IN EACH STAGE
Relationship building (2004/2005)	Meetings	Architect, students, NGO, community
Needs Assessment (2005)	Observations/surveys/ informal chats and structured questionnaires	Architect, students, NGO, community members
Design (2005)	Workshops	Architect, students, NGO, few community reps.
Testing of building components (2005)	Lab testing	Architect, local professionals
Construction process (2005 – 2006)	Co-construction	Architect, local professionals, skilled and unskilled labour from community, NGO
Official opening of the centre (2006)	Community event	Local administration rep., NGO, architect, community
Maintenance	Awareness programs initiated, supervised by NGO	Individual community member, NGO, architect

On the other hand, ethnographic investigation involved authors shifting between the roles of change agents and participant observers, thus moving inside and outside of the context. Within these roles there were a variety of elements recorded, as illustrated in Table 2.2. By simultaneously working with the two approaches of collaborative planning and ethnography, the authors steered into co-designing and co-constructing with the community, tapping into local issues, materials, tools, skills and labour. The Activity Centre was the tangible outcome produced 'by the people' and 'for the people' (Table 2.2).

Were the methods and methodology used able to build capacity in the people involved in the project? The next section defines capacity building within the context of this project, and assesses the success of this project through its capacity building outcomes.

Table 2.2 Methodology

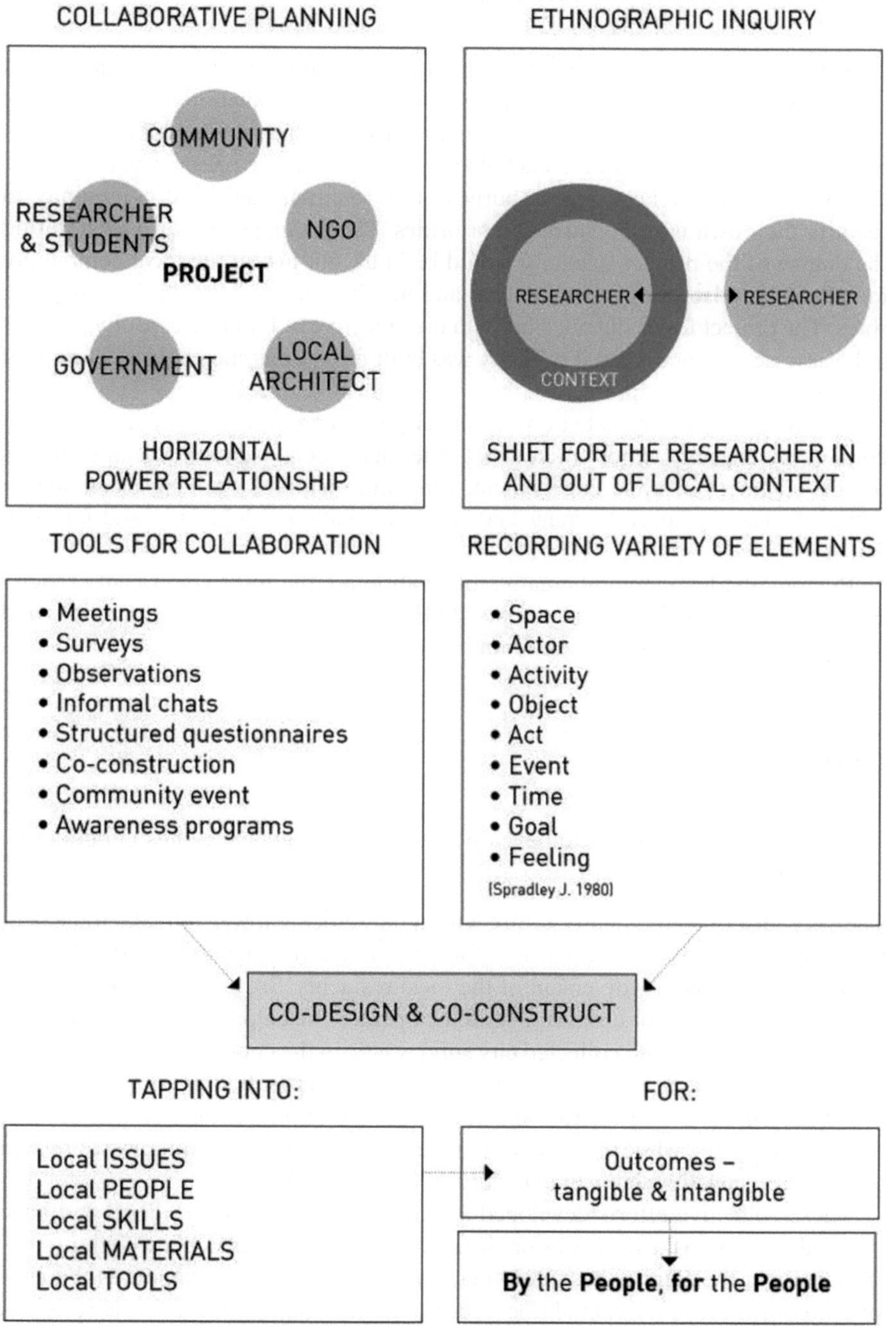

Capacity Building: What, and for Whom?

Capacity building is about empowering all partners in the project, so that, instead of upward answerability, downward or horizontal accountability is emphasized (Eade 2007). On this premise, and taking into account that our development project involves various contending stakeholders, we defined our roles as outsider professional and researchers, whose purpose is to facilitate the local community to identify their own issues, values and priorities. The presupposition was that during the course of the project, capacity would be built, not just in the local community members, but also in us as researchers and in other stakeholders in their different roles. The project for us did not start with the objective of building capacity, but was to be assessed as successful if capacity was built for various stakeholders during (or after) the course of the project.

How is a project evaluated on its capacity building outcomes? The concept of community capacity is not unlike what is termed 'social capital'. Social capital is the development of civic engagement, trust and reciprocity amongst community members, leading to co-learning and identity creation (McMurray and Clendon 2011).

Reciprocity, learning and creative innovation are the most significant capacity building outcomes (Innes and Booher 2003) and these are reflected tangibly and intangibly in the Activity Centre initiative.

Outcomes

1. Reciprocity is the ability to develop actors' interdependence, thus building recognition and identity.

The very idea of a community centre within the settlement was to make it the people's own, accessible to all. Critical for the formation of personal and collective identity is the direct involvement of the local residents. So, at Tekro, potters made roof-tiles and fly-ash bricks, semi-skilled carpenters made partitions and windows, and unemployed women collected city solid waste; in the construction work, housewives and old widows laboured, filling the glass and plastic bottles with fly-ash, and decorating and painting interior and exterior walls of the centre, using their local designs and skills; while children were given activities in their craft classes related to the centre development work (Pandya 2008) (Fig. 2.3).

These collective efforts developed a sense of ownership in residents and have inspired them to make decisions on their own for future extension and activity programming in the Centre, based on their changing needs. An example is the construction of a multi-religion shrine. During the installation of idols, the residents celebrated as each idol was offered by different faiths and by different construction agencies, such as contractors, masons, labourers. Yet another community initiative, introducing a gymnasium for the youth, was undertaken. Similarly, vocational train-

Fig. 2.3 Collective efforts resulting in a sense of ownership and identity. (Photos: Tiwari and Pandya)

ing programs have been introduced and computer training is conducted in the afternoons for youth to improve their employment prospects. Health camps (eye camp, gynaecology camp, malaria camp etc.) have been introduced, where all settlement dwellers are served. Another initiative has been the addition of a crèche which is run by elderly ladies of the settlement.

Thus, the community centre has been able to bring residents together and has empowered them to decide about issues regarding their own well-being.

2. Learning skills from each other is enabled because of strong relationships created between different project actors.

The hidden value of economic empowerment by transference of skill has been one of the most important outcomes for the community. The construction of the centre saw participation by volunteers on a weekly basis and used the services of local unskilled or skilled persons. Crate wood partitions were constructed by a local carpenter. Children participated in paper infill work in the door panelling during their art classes. Clay pots to be used as infill in the slab were procured from a neighbouring family working as potters. With guidance, the community learnt skills in reinterpreting waste materials and techniques for using them, and have been applying these skills in their own homes. An example has been the recycling of oil tin containers as door and window panelling, and this technique has been successfully adapted by the poorer segment of the settlement for their homes.

As well as these examples of capacity building for the community, we observed skills transference and learning by students and researchers. For students and researchers, the field visits and direct observation study were helpful in gaining a cross-cultural understanding and appreciation of the resourcefulness of the local community. Students learnt from the community while co-creating the architectural facility:

Fig. 2.4 Students interacting with the community and working on the project. (Photos: Tiwari)

> The India unit took us out of our comfort zones and taught us that the most important thing about design is who you are designing for. People are different all over the world and the rich culture of a place gives it meaning. The India trip taught us to respect that. (Michael Trees, journal entry)

For the last 2 days, students volunteered to provide hygiene instruction in primary schools as a thank you gesture to the community.

> Volunteering was really great … We used our Dettol wipes to clean these kids' hands, faces and fingernails. While I did this, Charlotte clipped fingernails and Margret brushed hair. It felt very rewarding. (Sam Austen, journal entry)
>
> Goodbye India … thankyou for teaching me never to give pity unduly … I will always remember to be grateful for what I have. (Melissa Ann Aroozoo, journal entry)

Besides the delivery of subject knowledge and exposure to international perspectives, such experiential projects help in building lifelong skills in students, such as adapting to and understanding a variety of situations and people while learning acceptance and tolerance of different lifestyles (Fig. 2.4).

Experiential learning draws from Stellje's Service Learning model (2008), linking theory to practice, and further contributes to democratic participation and power-sharing.

> It was also in this environment where the tutor became more like a work colleague in terms of feedback and bouncing ideas off each other. It was this relationship throughout the trip where the distribution of "power" between tutor and student was far more equal than in a normal classroom environment. I felt that this was crucial in the facilitation of learning and gaining knowledge and skills. (Student, Survey Monkey 2012)

3. Innovation results due to involvement and partnership with different actors and their knowledge is considered a means of reaching creativity.

Fig. 2.5 Innovative recycling of waste as building components. (Photos: Pandya)

Recycling of waste as building components has been a key innovative outcome and this has been possible due to the combined efforts of all stakeholders (Fig. 2.5).

The local cycle of waste collection and sorting was used to procure relevant materials. The project has been able to demonstrate the re-use of nearly twenty types of waste materials, and some building component prototypes were developed on the site by students or fabricators involved in evolving solutions based on the nature of waste (scrap) generated at that time. Industry participated by providing fly-ash from a thermal power station (producer of ash as waste) for the use of fly-ash bricks. Using municipal waste for building components reduces waste as pollution, and since the recycled building components are 20–30 % cheaper and of higher quality than conventional materials, they provide affordable and superior quality building alternatives for the urban poor. For this project, the materials were produced in small quantities as a demonstration; if they were mass produced in a decentralized way they would create further economy. The project of building components as a cottage industry potentially offers economic autonomy for the slum dwellers (Pandya 2008).

The above innovative outcomes were a result of the involvement of the wider community. The finances for construction came through donations from India and abroad, and thus it was the participation of the wider community that brought the project to fruition.

4. Spin offs and trickle effect

There were several additional positive results of the exercise, including: evolving a model to recycle municipal and domestic waste for reduced pollution and positive environmental impact; empowering women through economic opportunities by adding value and developing processes for transforming waste into building components; and creating a new palette of building components that perform cheaper and better than the conventional ones used by the urban poor, and make their homes affordable. There were some direct spinoffs. First it gave the local NGO and people lots of confidence; three additional centres have since been built in the settlement using a similar approach and ideals, as well as another one in a different settlement. The technique of using recycled oil tin containers for paneled doors has been adopted by the poorest communities within the settlement in their own homes, with pleasing adaptations like religious icons painted on them. The activities of the centre have expanded to include a women's centre, a co-operative store and a crèche minded by local grandmothers for slum children when their parents are at work. This has been the most self-sustained and convenient facility for the settlement. While the initial research and experimentation of building with recycled material was limited to slums, several applications generated keen interest from major developers as well as manufacturers in mass producing and emulating these methods in their projects. Fly ash brick manufacturing and plastic/rag rope making have produced popular mass products sold independently.

Challenges

The aim of collaborative planning was to have a horizontal relationship between all stakeholders, with everyone equal to each other. However, on the ground, things were different. All efforts to make this project successful were channelled through the NGO, which had a direct link to the community, researchers, architect and the government. Researchers and the architect had an indirect link to the community through the NGO, while the local government had no direct links to the community (Fig. 2.6). Thus, power resided with the NGO. The role of local government was negligible, perhaps because local government's participation would have raised bigger issues of landownership, infrastructure and eviction. Yet the involvement of local government bodies and other legislative bodies is essential to reorient the planning and financial measures so as to facilitate resident involvement in activities that meet their needs and aspirations.

Fig. 2.6 Stakeholder relationship

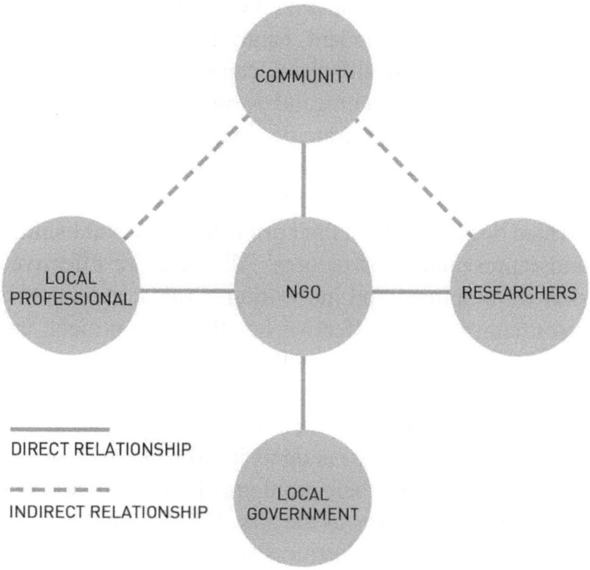

Conclusion

Significant outcomes of the Activity Centre project have been: a less polluting environment as a result of recycling of municipal waste; economic benefits to the community through the construction of affordable built forms; and skill transference and development for both students and community members. These have resulted from adopting the role of enthographer to identify opportunities within the local context, and facilitating alternative architectural technology that is low-capital intensive, and which uses and interprets local materials and skills, and involves the community and other stakeholders in collaborative decision making. Key aspects were the local context and the communal activity; the outcomes produced by the people had the potential to release artistic energy, bringing a deep sense of identity and ownership.

Active participation of local communities in all stages of the project is essential for its success: 'the people know their community and its issues; they have to live with the results, and can, want and have the right to participate' (Cities Alliance 2003, p. 21). This participatory strategy begins from a realistic needs assessment, and becomes a precondition of each stage of the project (Tiwari 2009). With their long-standing local associations, this is where the NGO's role as networkers becomes important.

People's participation depends not only on their direct presence in building but also a sense of involvement in many ways with its processes. For example, with the multi-religion shrine, all religious celebrations at the centre helped foster and cement strong bonds amongst the inhabitants of the neighbourhood. Similarly, their participation in and earnings from the centre's activities (as sub contractor, labourer, raw material provider or skilled worker during construction, and as teacher, security officer, manager/attendant, nanny for the children), along with socially beneficial programmes like health check camps and so on, helped make the community truly associated with the centre and owned by the residents as their neighbourhood centre.

Capacity building strategies include providing residents with skills to participate and identify their needs and issues, encouraging them to express their views on local issues before co-designing and co-constructing with them, letting them make informed decisions about their priorities, and establishing a wide range of participation and representation structures. All these are effective in encouraging reciprocity, skill development and innovation. Critical aspects for consideration are: adequate resourcing over an effective length of the project; enabling the residents to have genuine decision-making authority, so as to encourage social entrepreneurship; and sustaining motivation with 'quick-wins' to fight disappointment (Australian Housing and Urban Research Institute 2003).

Involvement of the residents in decision making empowers them to take responsibility and resolve their problems. Lefebvre's notion of 'right to the city' is about empowering the users of the space (Lefebvre 2002). In some ways, a rethinking of the role of designer is critical to recognizing the user as a potential resource and involving him/her as a positive design development tool in the process. This perception will help us transform houses into homes, shelters into places with character, clusters into community, enclosures into events, and buildings into living environments (Pandya 2008).

References

Adler, P. A., & Adler, P. (1997). *Membership roles in field research*. Thousand Oaks: Sage.

Australian Housing and Urban Research Institute. (2003). Achieving resident participation in community and urban renewal in Australia. *AHURI Research and Policy Bulletin, 19,* 1–4.

Barab, S. A., Thomas, M. K., Dodge, T., Squire, K., & Newell, M. (2004). Critical design ethnography: Designing for change. *Anthropology and Education Quarterly, 35*(2), 269–276.

Chaskin, R. (2001). Defining community capacity: A definitional framework and case studies from a comprehensive community initiative. *Urban Affairs Review, 36*(3), 291–323.

Cities Alliance: Cities without slums. (2003). Annual report. http://www.citiesalliance.org/ca/node/587. Accessed 30 Feb 2012.

Davoudi, S., & Healey, P. (1995). City challenge: Sustainable process or temporary gesture. *Environment & Planning C: Government and Policy, 13,* 79–95.

Debord, G. (1977). *The society of the spectacle* (trans: Perlman, F. & Supak, I.). library.nothingness.org. Accessed 20 April 2013.

Eade, D. (2007). Capacity building: Who builds whose capacity? *Development in Practice, 17*(4/5), 630–639. http://www.jstor.org/discover/10.2307/25548262?uid=3737536&uid=2129&uid=2&uid=70&uid=4&sid=21101387690133. Accessed 13 Feb 2013.

Gille, Z., & Riain, S. O. (1995). Global ethnography. *Annual Reviews Sociology, 28,* 271–295.

Hammersley, M. (1992). *What's wrong with ethnography? Methodological explorations*. London: Routledge.

Healey, P. (1997). *Collaborative planning: Shaping places in fragmented societies*. London: MacMillan.

Healey, P. (1999). Building institutional capacity through collaborative approaches to planning. *Environment and Planning A, 30*(9), 1531–1546.
Ife, J. (1995). *Community development: Creating community alternatives—vision, analysis and practice*. Sydney: Longman.
Innes, J., & Booher, D. (2003). The impact of collaborative planning on governance capacity. Institute of Urban and Regional Development Working Paper Series. Berkeley: University of California.
Kolko, J. (2011). Design for impact. http://www.slideshare.net/ixdsa/jon-kolko-design-for-impact. Accessed 20 March 2013.
Lewis, I. M. (1985). *Social anthropology in perspective*. Cambridge: Cambridge University Press.
Lefebvre, H. (2002). The right to the city. In G. Bridge & S. Watson (Eds.), *The Blackwell city reader* (pp. 367–374). UK: Blackwell.
Manav Sadhana. (n.d.). http://www.manavsadhna.org/subnode.aspx?snodeld=22&siteID=2&nodeId=7. Accessed 20 May 2013.
McMurray, A., & Clendon, J. (2011). *Community health and wellness: Primary health care in practice*. Australia: Elsevier.
Nilsson, M. (2000, April). *Organizational development as action research, ethnography, and beyond.* Paper presented at the Annual Meeting of the American Educational Research Association, New Orleans, LA.
Pandya, Y. (2008). Mass housing: A paradoxical idiom of identity within conformity: Modern housing solutions. *Architecture + Design, 25*(11), 46–52.
Reeves, S., Kuper, A., & Hodges, B. (2008). Qualitative research methodologies: Ethnography. *BMJ, 337*(7668), 512–514. http://dx.doi.org/10.1136/bmj.a Accessed 20 May 2013.
Sanders, E. B. N., & Stappers, P. J. (2008). Co-creation and the new landscapes of design. *CoDesign International Journal of CoCreation in Design and the Arts, 4*(1), 5–18.
Stelljes, A. (2008). *Service learning and community engagement.* NY: Cambria Press.
Tewdwr-Jones, M., & Allmendinger, P. (1998). Deconstructing communicative rationality: A critique of Habermasian communicative planning. *Environment and Planning A, 11*(30), 1975–1989.
Tiwari, R. (2009). Integrating the marginalized: Towards a connected city. In P. Felli (Ed.), *City Futures: Architetture Design Tecnologia per ill futuro delle citta* (pp. 139–153). Hoepli: Milan.
Trivedi, N., & Tiwari, R. (2010). Collaborative dialogue and action for home-based work issues in Indian slum settlements. *Reflections: Journal of Built Environment Research, 2*(1), 51–57.
World Architecture Community. (2009). Media release. http://www.worldarctitecture.org/world-buildings/world-buildings-detail.asp. Accessed 29 Jan 2012.

Chapter 3
Capacity Building and Community Development: A Community Dialogue on Equality in Rural Uganda

Debra Singh, May Lample, Mark Jones and Jaya Earnest

Abstract Community led capacity building applies and uses insights drawn from the community to address community concerns. Consciousness, reflection and sharing are critical steps in capacity building. The process of planning, action and reflection undertaken in this study ensured that an increasing depth of understanding was achieved by the participants. This cross-sectional, exploratory case study was conducted in 2011 in the post-conflict, transitional, developing nation of Uganda; the researchers documented changes in gender roles in courtship, marriage, education, health choices and agriculture. Twenty-one men and women in central Uganda participated in interviews and focus group discussions. The results reveal that women in rural Uganda have increasing decision-making power in the areas of agriculture, use of funds, educational choices for children, and partnerships. Women's health knowledge is highly valued not only in the family but also in the community. The participants shared that health is predominantly the domain of women, in terms of decision-making, patient care and providing advice. Men do not feel confident in dealing with sick children or family members, and defer to women. Lessons were learnt on the relationship between gender-related decision-making, empowerment and meaningful participation and how these inter-related concepts contribute to community development. As the study progressed, men and women presented opportunities for further capacity building to play a greater role in the family and in community based healthcare.

D. Singh (✉) · M. Lample
Kimanya-Ngeyo Foundation for Science and Education, Jinja, Uganda
e-mail: ahderom@gmail.com

M. Lample
e-mail: maymag9@gmail.com

M. Jones · J. Earnest
International Health, School of Nursing and Midwifery,
Curtin University, Perth, Australia
e-mail: Mark.Jones@curtin.edu.au

J. Earnest
e-mail: j.earnest@curtin.edu.au

R. Tiwari et al. (eds.), *M² Models and Methodologies for Community Engagement*,
DOI 10.1007/978-981-4585-11-8_3,

Keywords Women · Decision-making · Uganda · Participation · Empowerment · Community Development

Background

In many countries of the world, traditional patriarchal beliefs have historically translated to a lack of women's participation in decision-making in the family. This lack of decision-making power reflects traditional gender norms in African countries (Paek et al. 2008). A World Bank study in 2005 identified that women's lack of decision-making power over such issues as land, household assets, cash incomes, when and how often to have children, and healthcare, is one of the leading causes of poverty, and contributes to the prevalence of overburdened women and alcoholic men in Uganda. The Constitution of the Republic of Uganda (1995) gives women decision-making power at a policy level. This change has promoted equality and created awareness of the importance of gender equity for poverty eradication; however, the World Bank study found that there continues to be unequal decision-making in the family, to the detriment of both women and men (World Bank 2005; Fig. 3.1).

Wyrod (2008) interviewed men in his study of the relationship between men and women in Bwaise, a neighbourhood in Kampala, Uganda. He found many men who felt that women had taken away men's power; they felt that women were upsetting the natural social order, as men were more powerful than women. Even while some men agreed that women should have access to the same rights and services as men, they still had difficulty in relinquishing authority within their own homes. Some women agreed, believing that men have more power and authority but misused and abused this power. This understanding of power, that it is man's right and nature to control the actions of women, limits any opportunity for equal participation in decision-making in the home. This form of power sees men and women at odds with each, with one expecting to have power over the other (Allen 1998). In this situation, men feel it would disadvantage them for women to have equality.

The context for the possibility of transformation for women in Uganda is threefold. Firstly, the Ugandan Constitution unequivocally assigns and guarantees equal rights to both men and women; secondly, the government is committed to promoting the political participation of women; and thirdly, there is a growing involvement of women in the movement for cultural and religious rights. Since 1999, at a grassroots level, Uganda has begun to see a feminist movement, whereby women are taking action against inequality and mistreatment (Mbire-Barungi 1999; Fig. 3.2).

The changing landscape for women in Uganda is the result of participatory efforts by women at the community and grassroots level advocating for their own rights. The women's rights movement in Uganda has advocated for change in marriage laws and female land ownership, issues that directly target the sense of male authority and superiority (Wyrod 2008). Yet the change has been slow because of the refusal of men to accept gender equality and allow women to participate in

Fig. 3.1 Youth dancing at a community celebration, Jinja, Eastern Uganda. (Photo: Irene Basimike, used with permission)

Fig. 3.2 Children's class in Budondo, Eastern Uganda. (Photo: Irene Basimike, used with permission)

decision-making. Participation can be seen as a means of creating stronger ties within a community (World Health Organization 2008). Yet participation requires a desire by all members to see universal participation, as well as groups actively seeking their right to participate. For the second condition to happen, individuals must feel that they have the power to pursue what they need.

Community Development, Participation, Empowerment and Decision-Making

Community development involves the creation of improved societies through voluntary cooperation, participation and self-sustaining efforts that seek to contribute to the advancement of individuals within a certain region. The emphasis on community development allows the individuals themselves to shape their needs and action. Community development has two primary goals, to improve quality of life and to involve all people in the change process (Nikkhah and Redzuan 2009). Schafft and Greenwood (2003) agree, explaining that community development is an approach that allows individuals and communities to develop capacities focused on targeting community concerns.

In order for individuals and communities to become the protagonists for their own development, they must have the capacity to define, analyse and meet their own needs. The ability of a group to participate in its own development requires a wide range of capacities: namely, inclusive decision-making, ability to resolve conflicts and think systematically, and to understand problems and seek solutions (Institute for Studies in Global Prosperity 2008). In essence, it requires the full participation of all community members, based on the recognition that all have something worthwhile to contribute. Thus, community-based capacity building is both a prerequisite and an outcome of community development.

According to Nikkhah and Redzuan (2009) participation can be thought of as a collective effort by concerned individuals to bring their efforts and resources together, to carry out action they themselves have decided on. For individuals to have the opportunity to make decisions regarding their own lives and gain confidence, knowledge and skills, they must be allowed to participate in their own development. Participation seeks to empower individuals to take independent collective action in order to improve their own conditions (Economic and Social Commission for Asia and the Pacific 2009).

Through participation, individuals and groups begin to feel empowered as a result of their efforts and the practical experience of decision-making and becoming involved in consultation and taking action. These processes are transformative for the participants. It also leads to a greater consciousness of the forces that are taking place within the community as well as their ability to impact those forces (White 1996). Individual empowerment focuses on developing people's decision-making capacities, restoring their belief in their own autonomy (Kapiriri et al. 2003).

Participation does not necessarily lead to the sharing of power. The following questions need to be asked: who is participating? and: is the participation meaningful? Simply being present does not necessarily mean participation is truly taking place. Power dynamics in the community set the agenda for interests that may be addressed by participation. The interests aren't just there, they are chosen by those with the power in wider society (White 1996). Therefore, the power dynamics within a community must also change to allow truly representative participation, which alone cannot change conceptions of power. A space must be created in which participants can reflect and question current power dynamics within the community.

Therefore, empowerment should be considered not as dominating others, but as developing the power to act with others to create change (Mathie and Cunningham 2003). This shift brings a new understanding of the distribution of power in the community (Economic and Social Commission for Asia and the Pacific 2009). In this regard, individual participation necessarily impacts the motivation of the collective. As people see growing numbers of participants and how the actions of these participants impact their lives, it motivates them to engage in the process as well. The impact of participation is that the process empowers the participants themselves, as well as encouraging others to engage.

This brief background review has highlighted a number of inter-related processes that contribute to community development—participation, decision-making, and empowerment. All of these processes, when effectively undertaken, lead to capacity building within individuals and communities, enabling them to solve ever more complex problems facing their community. If capacity building does not occur, then long-term community development is unlikely.

A Research Case-Study: Women and Decision-making in Jinja and Buikwe

For this case-study, we wanted to understand and explore decision-making power in Ugandan families in Jinja and Buikwe. Furthermore, we wanted to examine if there was space for women to take greater decision-making initiatives. Participatory research methods were used that allow the participants themselves to identify issues in their communities. The value of using participatory methods is that they create trust between the researchers and participants. Trust is crucial in the examination of sensitive subject matter such as this, and this trust improves the quality and the quantity of the responses given by participants. Through this process, knowledge and experience is shared, and culturally appropriate measures are undertaken for the benefit of the community (Viswanathan et al. 2004). A space is then created for both men and women to discuss pressing problems and work together to eliminate them.

Questions were posed to participants by the lead researcher, the first author, who has lived in Jinja, Uganda for the past 6 years, and participants were given the opportunity to share and discuss their thoughts with other community members.

Individuals from various backgrounds came together in focus groups and reflected on the common practices in their communities. The participants were the protagonists, as they framed the evolution of the conversation. These focus groups created a space in which participants felt empowered to work together.

Community members invited to join the study were divided into two groups—those between the ages of 18–28 years and those over 45 years, and included both men and women living in rural areas. There were, therefore, two groups of men–younger and older–and similarly, two groups of women, with a total of four focus groups. These age groups were chosen because they would be in a position to represent the perspectives of two distinct generations. Twenty-two individuals—11 males and 11 females—participated in the study. The purpose of the conversations was to understand if the reality of women in the area reflected earlier literature about Ugandan women and decision-making, particularly related to health. The group process served not only to raise consciousness, but created a space for reflection and sharing amongst different sexes and generations when the data was analysed and shared with the focus group members and community members after the case study was complete. This study was approved by the Human Research Ethics Committee at Curtin University in Western Australia. The focus group discussions were facilitated by a team consisting of the first and second authors and community representatives. Analysis of the conversations highlighted several key themes of ongoing social change in Uganda and participation and power in the family.

Results and Analysis

Social Change, Participation and Power in Decision-making in the Family

The case study explored women's and men's views of how decision-making happens in the family. Participants discussed issues regarding responsibilities at home, family planning, health choices, agriculture and food. In their conversations, the participants described an increasing amount of conflict in the family (Fig. 3.3).

Women felt that men were not fulfilling their responsibilities, that there was a lack of trust in the families, and that women have been increasingly demanding support and are unwilling to put up with their husbands' bad behaviour:

> [In the] past they (men) were respectful but now they neglect paying school fees [and think it] is a burden and this is why children are misbehaving. Women are carrying 95 % of the responsibility and men are neglecting their responsibilities. (Focus group of women aged over 45 years)

These pressures may also result from high unemployment, the rising food costs, and an economic downturn (Hanson 2009). It was clear that older women were more confident in these decision-making areas, while young women and men were still

Fig. 3.3 Focus group participants, Budondo, Eastern Uganda. (Photo: Irene Basimike, used with permission)

trying to figure out their roles in relation to one another. For example, when asked who decides what the money will be used for and who controls the money, there was a difference in responses according to the age of the participants. The young women and men and older men felt that it was for the man to decide. The older women felt it was for the woman to decide:

> The mother will decide for the betterment of the family, they will decide what to bank, what to use for school fees and what to use for hospital fees. (Focus group of women aged over 45 years)
>
> The man decides. Buying food at home, buy a bicycle, alcohol, school fees. A man should not be asked how much is received and what it is used for. [If there are] grudges in the home, the man controls the money, if [the marriage is] in good condition then they decide together. (Focus group of men aged 18–28 years)
>
> [Both] mother and father have to make a decision [about how the money is spent]. Some men hide money, so the mother can also work and help the family. Women will hide money if the father is a drinker. Some fathers have many wives and he will take the money and come back even without food. [He] can use it on cards. Marriage has no law so they [the men] share their chance so they may have more than five or ten [wives]. (Focus group of women aged over 45 years)
>
> The man is free to buy alcohol, marry other women. Even when they are responsible, the man still decides how the money will be used. (Focus group of women aged 18–28 years)

When asked the question, 'what generally are considered the responsibilities of the mother and father at home?' one of the participants, reflecting on what she had known in her family and those of her friends, replied:

> [The woman] was meant to be a housewife, help at home, help children at home, teach them good behaviour and good morals. [The] man was meant to provide money for school fees/food. Most of the time clothing was the mother's responsibility [as well as] morals and behaviour. Mothers also helped with homework. For the father, [he paid the] bills. Father enforces that you have to do homework.... (Focus group of women aged 18–28 years)

In order to empower women to participate, space must be created for them to consult with men, to reflect on and evaluate their current social standings. Power has to evolve from the current understandings of power dynamics between women and men, which are at odds with cooperation, so that they can recognize that power is something everyone has, that cannot be lost or gained, but simply developed. Through the responses of the participants, it was clear that there was some blurring of gender roles that had previously been quite distinct:

> ...but the man [was] the vision bearer of the family— someone who stands for the family—education, health, decision-making. This is the past. Today both parties play a big role in the family. Now there are more permissive families—children are hardly there. (Focus group of women aged 18–28 years)
>
> In the past the husband would have a special seat and the mother and children would sit on the mat. Now in modern families they all sit together at the dining table. (Focus group of men aged 18–28 years)

Participation is crucial for the evolution of gender roles, because it enables individuals to develop a belief in their ability to make an impact on personal and social spaces (Schafft and Greenwood 2003). The beginnings of this were clear in the focus groups, as the younger women expressed a desire to continue to see the changing role of women:

> Girls should be put at the forefront, thought about as people, human beings that have flesh and blood like the boys. When it comes to jobs, girls can do anything. Girls have brains, girls can also be physical. Most of the women and girls are coming out to prove they are equal. They are putting aside culture —equal education, medication, making decisions and other kinds of things. Girls should not be left aside. [We need to] be taken seriously. Not be considered the inferior sex. (Focus group of women 18–28 years)

The thinking behind such conversations with people in the community is that individuals can begin to move beyond their own cultural influences, silence and self-blame, to redefine their social reality (Wallerstein and Berstein 1988). Rather than families suffering under the impact of negative social forces, equal decision-making means that the burden of responsibility does not fall on either men or women disproportionately.

> They (men and women) should share the work and share the money. If they (men) decide to not work then the money should be the women's. Women should be appreciated—thank you, a smile. The agricultural setting should change to encompass the old ways and the modern ways. Families should know what should be kept for consumption and sold.... Father and mother should work together without oppressing each other. Both should work together, cooperate and work together. (Focus group of women 18–28 years)

During our conversation, it became clear that health decisions in particular seem to be an area where women are considered to have expertise. When asked who makes the decisions on where and when to take a child for healthcare:

> When on good terms, it is the husband and wife [who decide where to seek healthcare]. When on not good terms it is the woman. She can take the child to the hospital. (Focus group of men 18–28 years)
>
> The mother will treat with tablets and will decide on what to do if the husband is not there. If he is there they will decide together, if he is drunk she will take the responsibility. (Focus group of women over 45 years)

When asked: 'If a woman decided to go to the clinic or take her child to a clinic without discussing this with her husband, would this cause a problem in the family?' the following responses were given:

> If she took the child to the wrong place then she will be blamed. It could cause problems if it costs a lot of money. Some ladies are not trustworthy at home—[they ask for] 5,000 shillings [approx $ 2] and keep the 3,000 shillings. Trust in the family is not there so there is always a problem. The wife is like the vice [chair person] in the family, so when the chair person is not there then she should take control. (Focus group of men 18–28 years)
>
> There is no problem. You need to communicate. If you don't have money and you need a lot of money for the treatment—why go to an expensive clinic. He will quarrel if she chooses the expensive one. Men do not always communicate well if they have the money or they don't. They need to communicate and they will tell them to take to government or a private clinic. (Focus group of men over 45 years)
>
> Children falling sick worries the parent. It is better to discuss with people who have once been parents. The woman is the main one who gives the care. The women take the responsibility. (Focus group of women 18–28 years)
>
> In the past it was the man. The woman decides now when and where [to take the child]. In the past a man could not leave home without leaving money. In health centres and private clinics you can get treatment and pay later. (Focus group of men over 45 years)

Discussion: The Way Forward

A possible outcome, as proposed by the community participants, would be an increased role for women as healthcare professionals, in particular as village health team members. Another capacity building opportunity the community identified was an increasing role for men in the family and community. The next phase of the process proposed, therefore, is to address both of these areas through capacity building initiatives at grassroots level, as a means of empowering both men and women in family and community life. Furthermore, as an ongoing study, the action and reflection process was seen as a means by which a deeper and more specific understanding of female participation in health-care and male participation in family life could be carried forward. Community development work is contingent upon the full participation of the people whom the work is intended to benefit. To this end, capacity building visualises individuals manifesting innate powers in action. In our conversations, we began to see this happening for the participants. As participants begin to understand their innate power, there will be a move towards translating that power into action (Institute for the Study of Global Prosperity 2008).

Although the basic principles of capacity building for community development are universal, the issues being identified and addressed are contextual. There is no community that can claim to be 'developed'. All communities are on a continuum of development, with its inherent challenges. Such challenging issues might include: family breakdown, mental illness and stress, drugs and alcohol, food security, equality of the sexes, education, poverty or ill-health, and land issues. According to the International Development Research Council, when development is looked at as 'change that improves the conditions of human well-being so that people can exercise meaningful choices for their own benefit and that of society' it places people at the centre of the process (Ryan 1995, p. v). The Institute for the Study of Global Prosperity (2008) defines development as capacity building:

> When viewed as capacity-building, development is concerned principally with the generation, application, and diffusion of knowledge... specifically... the world's inhabitants... must be engaged in applying knowledge to create well-being, thereby generating new knowledge and contributing in a substantial and meaningful way to human progress. (p. 7)

This case-study was set within the context of a rural community, and was an initial effort to create a space in which community members could reflect on gender dynamics in their families. The conversations held in the focus groups involved men or women in gender- and age-specific groups, with an emphasis on increasing participation as a means of developing the capacity to increase decision-making, and recognizing the ability of all members of the community to contribute to its advancement.

Community development, whatever the context, will face potential barriers that could block individuals' participation, including economic, social, cultural, attitudinal and motivational circumstances. Additionally, there may be participatory structures that, while sufficient for community participation, do not promote public involvement in planning and priority setting (Kapiriri et al. 2003). If barriers can be addressed by the community as a whole and there is true desire to collaborate, ways of overcoming the barriers, will, for the most part, be found. As more spaces for community members to come together are created, continued action can take place. An on-going planning, action and reflection process and continuum, illustrated in Fig. 3.4, draws on the strengths and skills of all community members and is vital for sustained long-term change.

Figure 3.4 illustrates the relationship between decision-making, participation and empowerment. These three aspects of community development are symbiotic and build on each other. The act of participation, whereby members of the community use their agency to contribute to their communities, leads them to feel more strongly empowered. They begin to believe in their own ability. This belief in turn encourages them to become more assertive, and to contribute to decision-making processes in their community. This in turn supports their participation, as they see their involvement yielding results. This framework can be used to facilitate continuous participation, empowerment and decision-making for members of the community. As this process takes hold at the grassroots, progress will be seen as individuals take ownership of their community development.

Fig. 3.4 A framework for community development

Conclusion

This case-study undertaken in rural Uganda illustrates the initial steps in the long-term process of addressing barriers to the full inclusion of women in decision-making processes within a rural community. Through initial dialogue, the community identified a number of challenges and opportunities which could be addressed. The lessons learned from this process have universal application, especially in rural communities. When people affected by the empowerment issues include women in decision-making and men and women have the opportunity to articulate their realities and to share their voices and concerns, new insights and learning can be generated. These insights can lead to community action to address and overcome the issues mentioned above. In the act of doing so, capacity and empowerment is built among the community members, to share and reflect on a new level of consciousness, and to address increasingly complex problems of living in the contemporary world.

Acknowledgements We would like to acknowledge Robinah, Julius and Emmanuel for their assistance in in the field during data collection. We would also like to thank the community members for giving so generously of their time. We are thankful to the Transcultural Health Improvement Programme, School of Nursing and Midwifery, Curtin University, for funding this study.

References

Allen, A. (1998). Rethinking power. *Hypatia, 13*(1), 21–40.

Economic and Social Commission for Asia and the Pacific. (2009). Regional trends, issues and practices in rural poverty reduction: Case studies on community participation. http://www.unescap.org/pdd/publications/poverty_and_development/trends_rural.pdf. Accessed 28 July 2013.

Hanson, H. (2009, November). Planning for instability in Kampala. Paper presented at the Association for African Studies annual meeting, New Orleans, LA.

Institute for the Study of Global Prosperity. (2008). Science, religion and development: Some initial considerations. www.globalprosperity.org/library. Accessed 28 July 2013.
Kapiriri, L., Norheim, O., & Heggenhougen, K. (2003). Public participation in health planning and priority setting at the district level in Uganda. *Health Policy and Planning, 18*(2), 205–213.
Mathie, A., & Cunningham, G. (2003). From clients to citizens: Asset-based community development as a strategy for community-driven development. *Development in Practice, 13*(5), 474–486.
Mbire-Barungi, B. (1999). Ugandan feminism: Political rhetoric or reality? *Women's Studies International Forum, 22*(4), 435–439.
Nikkhah, H., & Redzuan, M. (2009). Participation as a medium of empowerment in community development. *European Journal of Social Sciences, 11*(1), 170–176.
Paek, H., Lee, B., Salmon, C. T., & Witte, K. (2008). The contextual effects of gender norms, communication, and social capital on family planning behaviors in Uganda: A multilevel approach. *Health Education & Behaviour, 35*(4), 461–477.
Ryan, W. (1995). *Culture, spirituality, and economic development: Opening a dialogue*. IDRC. http://www.idrc.ca/EN/Resources/Publications/Pages/IDRCBookDetails.aspx?PublicationID=274. Accessed 28 July 2013.
Schafft, K., & Greenwood, D. (2003). Promises and dilemmas of participation: Action research, search conference methodology, and community development. *Community Development Society Journal, 34*(1), 18–35.
Viswanathan, M., Ammerman, A., Eng, E., Garlehner, G., Lohr, K. N., Griffith, D., et al. (2004). Community based participatory research: Assessing the evidence. *Evidence Report/Technology Assessment (Summary), 99,* 1–8.
Wallerstein, N., & Berstein, E. (1988). Empowerment education: Freire's ideas adapted to health education. *Health Education & Behavior, 15*(4), 379–394.
White, S. (1996). Depoliticizing development: The uses and abuses of participation. *Development in practice, 6*(1), 6–15.
World Bank. (2005). Uganda from periphery to center: A strategic country gender assessment. Office of the Sector Director poverty reduction and economic management Africa region (Report No. 30136-UG). http://www-wds.worldbank.org/servlet/WDSContentServer/WDSP/IB/2005/03/25/000090341_20050325085641/Rendered/PDF/301360UG.pdf. Accessed 28 July 2013.
World Health Organization. (2008). Closing the gap in a generation: Health equity through action on the social determinants of health. Commission on the Social Determinants of Health. Geneva: WHO Press. http://www.who.int/social_determinants/thecommission/finalreport/en/index.html. Accessed 28 July 2013.
Wyrod, R. (2008). Between women's rights and men's authority: Masculinity and shifting discourses of gender difference in urban Uganda. *Gender and Society, 22*(6), 199–823.

Chapter 4
Urban Neighbourhood Regeneration and Community Participation: An Unresolved Issue in the Barcelona Experience

Carmen Mendoza-Arroyo and Pere Vall-Casas

Abstract The urban neighbourhood regeneration in the Metropolitan Region of Barcelona (MRB), Spain, during the 1980s and early 90s provided significant good urban design outcomes, particularly its public open space production. However, this successful 'Barcelona model', based on an accurate morphology of the physical interventions, has masked the much less ambitious participatory processes. In general terms, Barcelona's experience shows a significant imbalance between the quality of both the urban design proposals and the participatory process of the urban renewal. The assessment of this experience solely from a physical perspective is controversial and for this reason an integrated approach is required. This chapter intends to contribute to this holistic analysis and delves into the correlation between physical outcomes and participatory processes in the urban neighbourhood regeneration of the MRB. According to this logic, it analyses the urban renewal tools in terms of their relationship between the participatory process and the urban design quality. Specifically, we state that the inclusiveness of formal participatory frameworks doesn't necessarily guarantee the higher civic potential of the physical outcome.

Keywords Neighbourhood regeneration · Participatory process · Urban renewal tools

C. Mendoza-Arroyo (✉) · P. Vall-Casas
Department of Urban Design and Regional Planning, School of Architecture, Universitat Internacional de Catalunya, Barcelona, Spain
e-mail: cmendoza@uic.es

P. Vall-Casas
e-mail: perevall@uic.es

R. Tiwari et al. (eds.), *M² Models and Methodologies for Community Engagement,*
DOI 10.1007/978-981-4585-11-8_4, © Springer Science+Business Media Singapore 2014

Introduction

Urban neighbourhood regeneration in the Metropolitan Region of Barcelona (MRB), Spain, has provided significant physical outcomes. In particular, Barcelona's urban renewal in the 1980s and early 1990s brought changes to the public open space through urban projects (Solà-Morales 1987) or planning projects (Marshall 2004), and has become an international standard of good urban design[1] (Busquets 2004; Bohigas 1985; Ajuntament de Barcelona 1983, 1987, 1996; Montaner 1999). However, this assessment of a successful 'Barcelona model' (Monclus 2003; Blanco 2009), based on an accurate morphology of the physical interventions, has masked the much less ambitious participatory processes.

In general terms, Barcelona's experience shows a significant imbalance between the quality of the urban design proposals and the participatory process of the urban renewal. The assessment of this experience solely from a physical perspective is inadequate, and a more balanced result requires an integrated approach. This chapter contributes to a holistic assessment, and investigates the correlation between physical outcomes and participatory processes in the urban neighbourhood regeneration of the MRB during the past thirty years. Specifically, we argue that the inclusion of formal participatory frameworks doesn't necessarily guarantee the higher civic potential of the physical outcome.

The first section of the chapter presents the changes in urban renewal tools aligned with political changes initiated by the first democratic municipal governments in the 1980s. The second analyses three main tools applied in the MRB from the 1980s until now. The third compares the three tools and analyses the effects of the formal participatory process on the civic potential of the physical proposals. Finally, the fourth part develops conclusions that may guide the implementation of further urban renewal tools in terms of the relationship between the participatory process and the desired urban design quality. The chapter is based on the authors' research developed at the Centre of Urban Design and Planning at ESARQ-UIC (www.intermediatelandscapes.com) as well as on supervision of research on urban regeneration and participation, and the interpretation of literature on participatory and urban renewal tools (Arnstein 1969; Blanco et al. 2011; Blanco and Brugue 2010; Bohigas 1985; Busquets 2004; Font and Galais 2011; Fung 2006; Gomà and Rosetti 1998; Martí-Costa and Cruz 2009; Mongil 2010; Nel·lo 2008, 2009, 2010, 2011, 2012).

[1] The city obtained the 'Prince of Wales in Urban Design' award in 1990, the 'Urban Public Spaces of Barcelona 1981–1987' Cambridge award in 1991, and in 1998 the Royal Institute of British Architects (RIBA) of London awarded the city the 'gold medal' in recognition of its outstanding urban works.

Democratic Municipal Governments: Framing the MRB's Urban Regeneration Tools

During the past thirty years, community participation in urban neighbourhood regeneration in the MRB has shifted from a non-formal role to the implementation of formal frameworks for citizen participation (Ajuntament de Barcelona 2002). However, although the formal framework for participatory process has been progressively established and widely assumed during the democratic period, the real effect of this framework in public decision-making is uncertain. The urban renewal of the MRB has evolved from a political agenda motivated by social demands during the 1970s and 1980s towards recent privately-led strategies. In order to understand this evolution, we focus on three different periods that set the framework for significant neighbourhood regeneration tools in the MRB.

Regeneration in Political Transition (1970s–mid 1980s)

To understand the essence of the participatory process in the planning strategies that have guided Barcelona's urban regeneration during the 1980s, one must look back to the 1970s. During the last stage of General Franco's dictatorship (1939–1975), urban interventions were top down, excluding community participation; examples of these were the inner city renewal policies of the 1976 Barcelona Metropolitan Plan (BMP),[2] which included thoroughfare interventions that demolished existing buildings and thus destroyed their social fabric. Due to these non-consensual processes, civil society reacted, and strong organizations emerged in the historic neighbourhoods of Barcelona, such as: the Neighbourhood Association of the Fifth District in the Raval Neighbourhood, and the Neighbourhood Association of the Historic Centre (Blanco et al. 2011). Together they created a socio-cultural political convergence, which nurtured the political programmes of the democratic parties and guided the 'democratic urbanism' of Barcelona (Borja 2010).

The strength of this popular movement relied on the strong correlation of three factors: (1) the increasing social and urban demands highlighting lack of social facilities, public spaces, and housing betterment; (2) criticism by intellectuals and professionals of the speculative urban development of the regime; and (3) local democratic demands. In this context, the first task the new municipal government had to face was to adapt the existing planning tools and the legal and participatory framework to fit these urgent needs. According to this, two objectives became essential: (1) the implementation of a democratic municipal governance structure; and (2) the development of urban renewal projects with the direct participation of social movements. The first objective was accomplished with the decentralization of the municipal government in ten districts of Barcelona (Blanco et al. 2011). The second objective was not totally fulfilled, due to the inadequate participatory mode

[2] The current Metropolitan Plan of Barcelona: *Plan General Metropolitano de Barcelona*, 1976.

inherited from the 1976 Barcelona Metropolitan Plan, which was based on two periods for public information and consultations, at the initial and final stages of the plan. Thus, the role of citizens in decision-making was limited to 'informative' and 'consultative' participation (Arnstein 1969). In sum, the level of inclusiveness of the formal participatory process was low, and the expected enhancement of citizen involvement with the arrival of the democratic governments did not occur. Towards the mid 1980s, the neighbourhood associations weakened, due to many former leaders becoming public employees and local councillors (Busquets 2004; Blanco and Brugue 2010).

However, despite these limitations, the new municipal agents and the neighbourhood movements led the design and development of the Special Plan for Inner-city Renewal (SPIR)[3] during the first democratic legislature (1979–1983), and their approval during the second legislature (1983–1987) (Gomà and Rosetti 1998; Bonet i Marti 2012). The aim of the SPIRs was to provide a global approach for neighbourhood upgrading and, at the same time, include diverse specific urban projects. The criteria for the selection of the neighbourhoods for SPIR were three: (1) neighbourhoods with an urgent need for global and yet particular urban betterment solutions; (2) neighbourhoods that would give high performance to the plans and projects; and (3) neighbourhoods with strong community organizations (Bohigas 1985).

In the case of the historic centre, Casc Antic of Barcelona, the design of the SPIR was based on what was called the Popular Plan of the Casc Antic. This plan was designed by the neighbourhood associations and independent professionals during the democratic transition, and incorporated a diagnosis of the area's social and physical needs as well as alternatives to the urban development model of Franco's regime. We shall analyse this case in the next section in terms of its physical interventions and participatory process.

Public-Private Regeneration (1983–1996)

With the second municipal government (1983), the urban renewal of the city centre demanded large investments, and public-private cooperation was required in order to facilitate and speed up urban renewal. For this purpose, a new ad hoc institutional structure called Integrated Rehabilitation Areas (ARI)[4] was implemented (Abella 2004; Cabrera 1998; Blanco et al. 2011). Citizen participation was incorporated in the ARI by law, and included neighbourhood representatives in the management committee responsible for supervising the economic, executive and regulatory process of the rehabilitation. The committee was constituted by representatives of the different tiers of government—regional, municipal and local, as well as the chamber of commerce, and included a representative of each of the four neighbourhood asso-

[3] *Plan Especial de Reforma Interior* (PERI).

[4] *Área de Rehabilitación Integral* (ARI).

ciations of the historic centre (Raval, Casc Antic, Barri Gòtic and Barceloneta). This vast representation was complemented in 1988 by the creation of a private-public management company PROCIVESA, later called FOCIVESA, which broadened the scope by incorporating the main association of retailers of the historic centre as shareholders. Finally, this company was dissolved in 2006 when it was considered that the big urban regeneration projects of the city were concluded.

In sum, the historic centre committee displayed a highly inclusive formal participatory process. However, it was unable to conceal the shift from state-led urban management to public-private cooperation, and the associated incorporation of subsidies for private rehabilitation of housing without implementing measures for social or economic development (Blanco et al. 2011, p. 332).

Public-Public Regeneration (2004–2010)

At the end of the 1990s, the tools and measures for the regeneration of Barcelona's historic centre won international recognition and a UN-Habitat's 1998 'Best Practice award'. Despite this success, many neighbourhoods in the MRB were still facing urban deficits and spatial segregation. To face these physical, demographic and socio-economic problems, the Catalan regional government approved the Neighbourhood Act.[5] The five-year period (2004–2010) in which this Act was implemented has been extensively studied (Nel·lo 2008, 2009, 2010, 2011; Muñoz 2006; Mongil 2010; Marti-Costa and Parés 2009). The singularities of the law were: (1) its crosscutting approach addressing social, economic, physical and environmental problems; and (2) its public-public management with Catalan and municipal governments as sole financial backers. The regional government established an investment fund and annually held open calls for municipalities to participate in a contest format, presenting Comprehensive Intervention Plans (CIP) for one of their neighbourhoods. If selected, the regional government granted the funding for 50–75 % of the projects of the CIP, and the remainder was financed by the municipal government. The selection criteria were based on 16 indicators, including social, physical, environmental and demographic indicators of the neighbourhood. The severity of the social and physical deficits was highly relevant for selection.

In order to analyse in depth the regenerative process involved during the democratic regime, the next section focuses on the three main betterment tools: (1) the Special Plan for Inner-city Renewal (SPIR) (1980–1990); (2) the Integrated Rehabilitation Area (ARI) (1983–1996), both implemented by the municipal governments; and (3), the Comprehensive Intervention Plan (CIP) of the Neighbourhood Act (2004–2010), propelled by the regional government.

[5] *Act 2/2004 (Catalan Government) Llei 2/2004, de 4 de juny, de millora de barris, àrees urbanes i viles que requereixen una atenció especial, Llei de Barris.* 'Act 2/2004 on neighbourhoods, urban areas and towns requiring special attention' known as the Neighbourhood Act.

Spir, Ari and Cip: Regenerative Tools for the Barcelona Metropolitan Region

The analysis of the three urban renewal tools will focus on two key aspects aligning to the objective of our chapter: the participatory process involved, and the physical proposals developed. Following, the objective is to describe the tools in order to later compare them.

The SPIR and the Civic-led Regeneration of Barcelona's Historic Centre

Participatory Process

The nature of the SPIR's process was inclusive, as each one of the ten districts of Barcelona was provided with a council in order to guarantee formal citizen participation. The councils met with community leaders prior to and after the plan's elaboration, applying the process of consultation established in the 1976 Barcelona Metropolitan Plan (BMP). The specific town planning needs were negotiated with the community movement in each neighbourhood, particularly in those with a greater history of citizen participation (Blanco et al. 2011, p. 332). In this respect, the formal participation was consultative in nature and weak in the decision-making process. However, due to the high involvement of the social movements in the development of the Popular Plan of the Casc Antic which the SPIR of the historic centre was based on, the urban projects of the SPIR transcended the building scale and framed a network of significant civic places.

Physical Proposal

The SPIR aimed at specific regeneration projects on parks, squares, streets and public facilities. The objective was to create urban references and interrelations by finding an equilibrium and correspondence between the social life and the neighborhood's physical structure through a systemic approach. The resultant 'civic network' (Vall 2007; Vall et al. 2011) became the driver of a betterment strategy referred to by Oriol Bohigas, chief architect of the city council, as '*Fer ciutat*', meaning to 'construct urbanity' by fulfilling the everyday needs of public space and social facilities for a quality urban life (Bohigas 1985).

Fig. 4.1 An example of a commercial piecemeal intervention of this period, the Maremagnum mall in the Port Vell, Barcelona, Spain. (Photo: David Lladó-Porta)

The Integrated Rehabilitation Area (ARI): a Change in Institutional Governance for the Rehabilitation of the Historic Centre

Participatory Process

Although the law of the ARI supported local participation by making it mandatory to widen the scope of representation, the role of the participants was limited to information and deliberation on institutional and social agreements. In sum, the major advance regarding participation was on inclusiveness (Font and Galais 2011), with a richer representation of societal sectors and stakeholders. However, stakeholders participated in participation (Arnstein 1969) in a planning stage where there was little opportunity to influence the decision-making, and the objective of participation was limited to achieve institutional and social agreement over the disputes concerning expropriations, compensations and reallocations.

Physical Proposal

The urban projects developed at the end of this period tended towards a piecemeal approach due to the public-private urban management involved. Isolated open space and commercial and housing projects were favoured, as a result of the rising interest in touristic and private investment enhanced by the betterment of the historic centre. Figure 4.1 shows a privately owned commercial mall in the Port Vell of Barcelona, which was a single concept building that was not integrated as a component of the whole historic centre. Due to this approach, despite its attractive and high quality design, it does not contribute to the residents' quality of life; on the contrary it invades the openness of the existing waterfront, blocking the view for the neighbours of the city centre and modifying Barcelona's skyline from the sea (Fig. 4.1).

Moreover, the urban betterment gave rise to increases in land value, leading to social exclusion and gentrification of the local community (Blanco et al. 2011). At this point, although the ARISs gave incentive to inclusive participation, they offered residents little opportunity to react against these negative consequences. In this sense, Borja (2010) states that when urban interventions submit to piecemeal architecture, it is the city that loses. Without doubt, the 1990s gradually shifted away from the systemic approach of the first democratic councils towards iconic, isolated urban projects.

The Catalan Regional Government Neighbourhood Regeneration: the Comprehensive Intervention Plan (CIP)

Participatory Process

The mechanism for citizen participation introduced by the Neighbourhood Act was a Committee for Evaluation and Assessment of the CIP. Its functions were to establish the coordination and information mechanisms, such as a semester report assessing the interventions and a final evaluation. The municipal government, along with economic, social and community agents integrated the committee. Although participation was considered an important part of the plan, the committee was established once the resources were assigned and the plans developed, consequently limiting the opportunity for citizen decision-making. However, in some neighbourhoods where community participation previously existed, and the municipal will enhanced this practice, community-based decision-making was achieved (Martí-Costa and Parés 2009).

At this point, to delve further into the nature of this tool, we briefly analyse the CIPs of the neighbourhoods of the Barriada Nova in the city of Canovelles and the Barri Congost in the city of Granollers,[6] located in the BMR. We selected these CIPs because of our authorship of both and the good understanding of the regenerative process acquired.

The Barriada Nova and Barri Congost neighbourhoods were once two separate suburbs from their respective historic cores, Canovelles and Granollers, but today they form a dense urban continuum. At the end of Franco's dictatorship, both neighbourhoods faced strong urban deficits, such as low quality housing and public in-

[6] Comprehensive Integral Plan (CIP) of the Barriada Nova, Canovelles (2005) and Barri Congost, Granollers (2006). Authors: Carmen Mendoza-Arroyo and David Lladó-Porta co-principals DA-Carquitectura, Rehabilitació i Urbanisme, S.L.P.

frastructure, and a lack of social facilities and open spaces. Active neighbourhood associations,[7] which acquired stability during the first democratic municipal governments, contested these issues, and many leaders were elected as local councilors. This strong and long-term connection between the neighbourhood associations and their respective councils promoted participatory actions during the development of both CIP's urban projects. Actually, when the CIP development began, there were programmed interventions propelled by the municipal team, which were able to continue with the economic support of the Neighbourhood Act.

The technical team formed by our firm and the Municipal Urban Projects Office of Granollers and Canovelles were sole developers of both CIPs. The participation took place in a consultative manner, and the Urban Project's office held unilateral meetings with the Committee for Evaluation and Assessment of each neighbourhood. Two open presentations were held at the Municipal Library, in which our team and the representative of the Urban Projects office presented the CIP and answered questions and doubts. However, despite the good intention of the municipal team, this voluntary process of participation was not enough to balance the absence of a proper participatory method involving the community in decision-making in all its phases: diagnosis, design, implementation and assessment.

Physical Proposal

Prior to the Neighbourhood Act, the betterment projects both councils developed had a piecemeal approach due to their limited budget. The shift toward a more comprehensive urban renewal began in the year 2000 with the revision of their Municipal Town Plan (POUM), in which a close consultation process between the neighborhood association leaders and the councils took place. On the other hand, as part of the POUM, the councils of both neighbourhoods commissioned us to develop a thorough analysis and betterment proposal for the whole neighborhood, which later became the framework of the CIP.

The objective of the CIP was to connect in between public and private spaces, and by doing so, establish specific guidelines for diverse architectural components ensuring urban coherence.

Figures 4.2 and 4.3 map all the urban repair interventions of open spaces linked to social facilities. The resultant network increased the social cohesion of the neighbourhood and the urban content of the urban projects, and generated a catalytic betterment effect.

This comprehensive 'neighbourhood approach' (Lowndes and Sullivan 2008) incorporated feedback from the POUM's participatory process, such as the need for creating spaces for community interaction. Likewise, during the design of the open spaces that Fig. 4.4 shows, we incorporated suggestions provided by neighbour-

[7] The Associació de Veïns Quinze Regions in Canovelles, and Associació de Veïns de Congost in Granollers

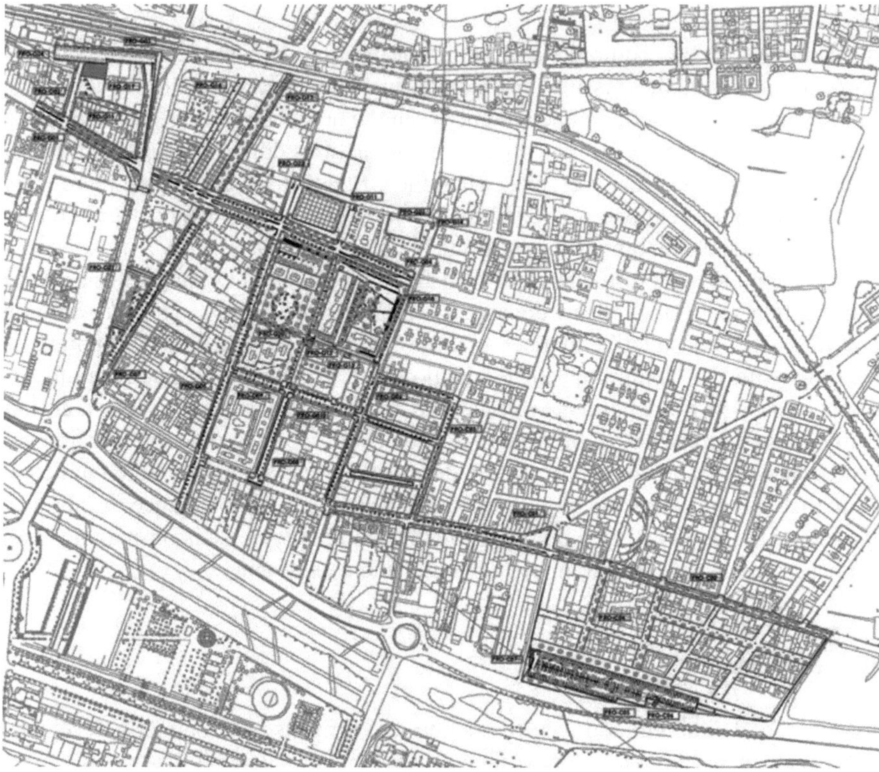

Fig. 4.2 Interconnected urban interventions in order to propel systemic regeneration. CIP Barriada Nova, Canovelles, MRB, Spain. (Figure: C. Mendoza-Arroyo and D. Lladó-Porta)

hood leaders to the municipal technicians through the Committee for Evaluation and Assessment, such as space for sports and parking.

The use of public architectural projects as part of a structural betterment creates landmarks (Lynch 1960) for the community, and shows that architecture- and design-based spatial planning can be an extremely strong medium to transcend physical boundaries when there is community support.

Integrated Assessment of Physical and Social Components

In order to address the correlation between the quality of the physical proposals and the inclusiveness of the formal participatory processes for each of the three renewal tools introduced in the previous section, we start by defining both concepts.

Firstly, the quality of the physical proposal is measured by considering its contribution to building a consistent system of public spaces. Hence, the best scenario, exemplified in Fig. 4.5, would be place-based urban projects conceived as part of a

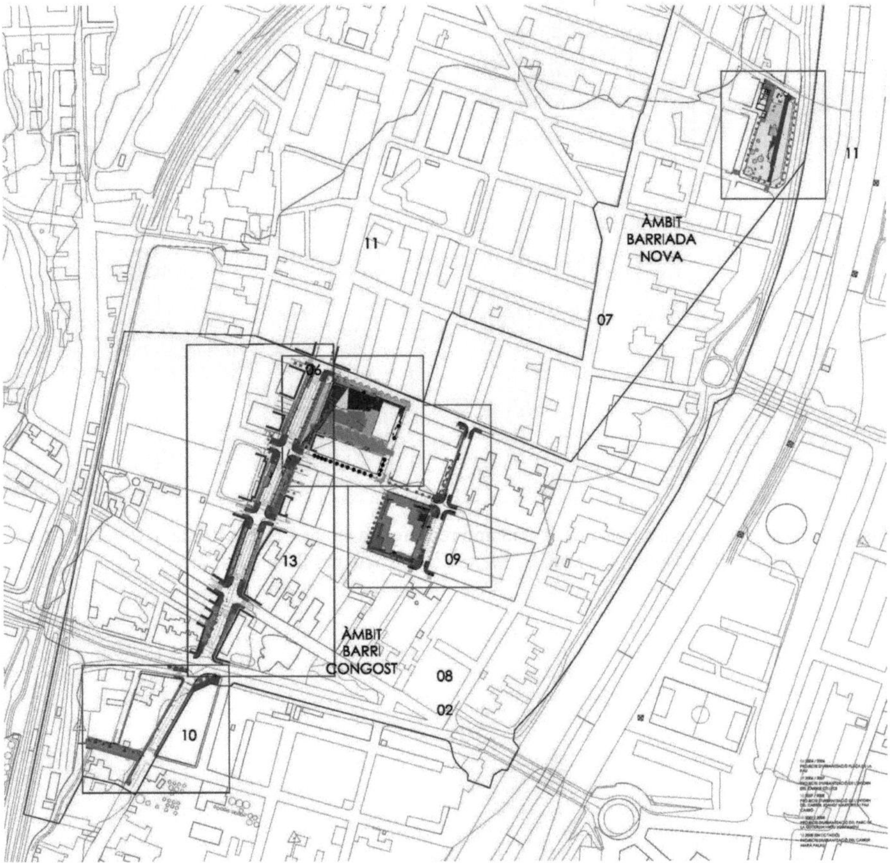

Fig. 4.3 Specific urban interventions addressed to create new public spaces and facilities. CIP's Barri Congost and Barriada Nova, MRB, Spain. (Figure: C. Mendoza-Arroyo and D. Lladó-Porta)

civic network of connected, significant places such as parks, plazas and boulevards (Vall 2007; Vall et al. 2011). We consider that this systemic vision boosts the integration of communities and enhances the existing social ties among citizens of diverse profile. On the other hand, the piecemeal approach relies on separate projects as designed pieces, with no care taken to fit them into an overall syntax, thus failing to acknowledge the opportunities for ensuring urban coherence in conjunction with the existing socio-spatial relations.

Secondly, the inclusiveness of the formal participatory process is measured through two indicators: (1) the type, and (2) the moment of participation. The type of participatory process we consider to be optimal is partnership participation according to Arnstein's (1969) typology,[8] for it enables citizens to take part in the

[8] Sherry Arnstein (1969) develops a typology of participation through a ladder of empowerment with eight rungs: manipulation, therapy, informing, consultation, placation, partnership, delegated power, and citizen control.

Fig. 4.4 New public space Plaça de la Pau and redesigned street Carrer Girona. CIP's Barri Congost, Granollers MRB, Spain. (Figure: C. Mendoza-Arroyo and D. Lladó-Porta)

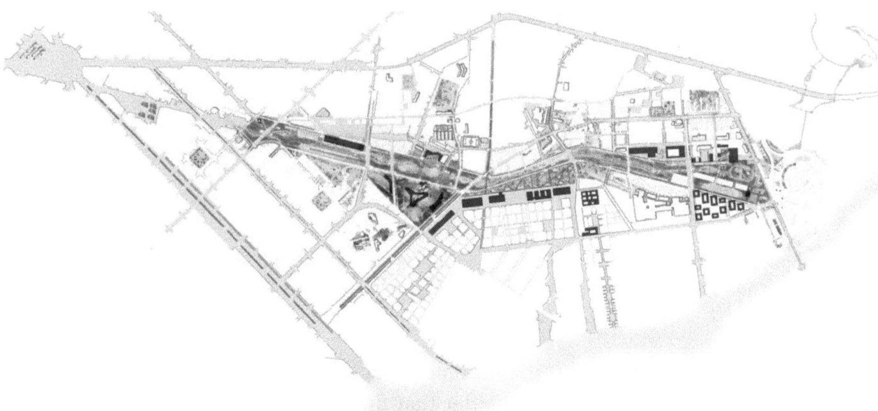

Fig. 4.5 Civic grid of significant axis and places for Sagrera Station area, Barcelona, Spain. (Figure: C. Mendoza-Arroyo and ESARQ-UIC final degree students 2012–2013)

decision-making and allows negotiation and trade-offs with the traditional power-holders. On the other hand, the informing and consultation types allow less connection between the participatory process and the plan's outcome.

Regarding the moment participation should take place, we believe the optimum is for it to happen in a crosscutting way: before, during and after the plan's development. The moment when the participatory process occurs correlates with the citizen's capacity to influence the process. If the participation occurs before, it may imply co-design and co-decision (Font and Galais 2011) of citizens' suggestions

Table 4.1 Participatory process and physical proposal

REGENERATION TOOL	PARTICIPATORY PROCES			PHYSICAL PROPOSAL
	Type of Participation	Moment of Participation	IFPP*	Systemic Oriented (S.O)/Piecemeal Oriented (P.O)
Special Plan for Inner – city Renewal (SPIR)	Informing and Consultative Two periods of public consultation	Before Plan Approval	Low	S.O
Integrated Rehabilitation Area (ARI)	Informing and Consultative Management Committee	Before Plan Approval	Medium	P.O
Comprehensive Intervention Plan (CIP)	Consultative Committee for Evaluation and assessment	After Plan Approval	Low	P.O

and ideas. On the contrary, if participation only occurs after the plan is developed, there is no real possibility of influencing the plan. Fung (2006) states that citizen empowerment occurs when the participatory process is linked to policies, allowing participation to take place during the whole process. Therefore, the criteria of type and moment together allow us to evaluate the level of Inclusiveness of the Formal Participatory Process (IFPP) for each urban renewal tool.

In order to explore the effect of the quality of the participatory processes in the quality of the physical proposals, Table 4.1 integrates both components and allows us to understand the particular interactions developed in SPIR, ARI and CIP, which are further elaborated below.

SPIR The SPIRs contributed the most systemic oriented physical outcome, despite the fact that this period featured the less inclusive formal participatory process. This apparent contradiction is resolved when we acknowledge that the SPIRs were based on the diagnosis and proposals previously elaborated by the neighborhood associations, with the involvement of the civic forces as ideologists and co-organizers of the regeneration process. For this reason, although the IFPP was low, a real collective decision-making guaranteed a higher civic quality of the diverse urban projects of the SPIRs.

ARI On the contrary, the ARIs, with the more inclusive participatory approach based on the mandatory management committees, generated a piecemeal approach, due to the lack of real civic involvement in the decision-making. A possible reason behind this fact is that by then, the social movements that sprang up at the beginning of the democracy had lost their critical attitude and social base.

CIP The participatory process in the design of the CIPs was not at all influential on the physical outcome, because consultation occurred once the plans had been developed. On the other hand, the funding selection process generated uncertainty, so that a long process of participation was seen as a burden to the plan development. Unfortunately, the commitment to the goal of the Act wasn't achieved, that is, an integrated physical and social approach remained a personal option of the professionals and municipal government involved.

Conclusions

In our view, correlation between the quality of the formal participatory process and the quality of the urban renewal plan is a relevant issue. According to this logic, the comparative analysis of the regenerative urban tools of the MRB analysed has shown that less inclusive formal participatory processes with high levels of influence in decision-making enhance systemically oriented physical proposals. On the other hand, more inclusive formal participatory frameworks may be less effective in their influence on decision-making and building civic networks. The evidence reviewed in this chapter supports our hypothesis that inclusiveness in the formal participatory processes doesn't necessarily guarantee a better civic quality of the physical outcome. We offer two main conclusions to be considered in relation to assessing Barcelona's urban regeneration, which could be useful for further implementation of renewal tools:

1. Formal consultative participatory frameworks may conceal the real extent of the participatory process. Less formal processes, when grounded on stronger community involvement, have proven to be more effective for enhancing systemically oriented renewals.
2. The quality and civic content of the physical transformation propelled by the renewal plan should be more related to its capacity in generating a public space system accessible to all citizens, and less to the iconic image of isolated urban projects.

Both conclusions reinforce the contribution of authors that criticize the threat of an empty formal participation. Among others, Arnstein (1969, p. 1) stresses that there is a critical difference between going through the empty ritual of participation and having enough impact to affect the outcome of the processes; and Fung (2006) states that mechanisms of participation are mostly designed to address particular problems in specific conditions, therefore responding to specific problems rather than to the essence of participation as such. It should also be added that we firmly believe core participatory processes must build on grassroots compromises and partnerships between citizens, stakeholders and administration. Definitively, the involvement of the civic forces as co-designers and co-organizers of the regeneration proposals remains essential to guarantee the civic quality of the physical outcome.

Acknowledgements The authors would like to thank Nacho Vallhonrat for his help with the formatting of the images of the chapter, and the editors for their thorough revisions and inspiring suggestions.

References

Abella, M. (2004). *El centre històric reviscolat*. Barcelona: Aula Barcelona.

Ajuntament de Barcelona. (1983). *Plans i projectes per a Barcelona*. Barcelona: Ajuntament de Barcelona.

Ajuntament de Barcelona. (1987). *Urbanisme a Barcelona: Plans cap al 92*. Barcelona: Ajuntament de Barcelona.

Ajuntament de Barcelona. (1996). *Barcelona: La segona renovació*. Barcelona: Ajuntament de Barcelona.

Ajuntament de Barcelona. (2002). *Normas reguladoras de participación ciutadana*. Barcelona: Ajuntament de Barcelona.

Arnstein, S. (1969). A ladder of citizen participation. *Journal of the American Institute of Planners, 35*(4), 216–224.

Blanco, I. (2009). Gobernanza urbana y políticas de regeneración: el caso de Barcelona. *Revista Española de Ciencia Política, 20*(4), 125–146.

Blanco, I., & Brugue, Q. (2010). *Urban governance networks and democracy: Which networks? What democracy? Why bother?* Paper presented at the 2010 Political Studies Association Conference, Edinburgh, Scotland.

Blanco, I., Bonet, J., & Walliser, A. (2011). Urban governance and regeneration policies in historic city centres: Madrid and Barcelona. *Urban Research and Practice, 4*(3), 75–99.

Bohigas, O. (1985). *Reconstrucció de Barcelona*. Barcelona: Edicions 62.

Bonet i Marti, J. (2012). El territorio como espacio de radicalización democrática. Una aproximación crítica a los procesos de participación ciudadana en las políticas urbanas de Madrid y Barcelona. *Athenea Digital*, 3(1), 15–28.

Borja, J. (2010). *LLums i ombres de l'urbanisme de Barcelona*. Barcelona: Editorial Empúries.

Busquets, J. (2004). *Barcelona la construcción urbanística de una ciudad compacta*. Barcelona: Ediciones del Serbal.

Cabrera, P. (1998). La transformació urbana de la Ciutat Vella 1988–1998. *Barcelona i Societat, 9,* 14–31.

Font, J., & Galais, C. (2011). The qualities of local participation: The explanatory role of ideology, external support and civil society as organizer. *International Journal of Urban and Regional Research, 35*(5), 932–948.

Fung, A. (2006). Varieties of participation in complex governance. *Public Administration Review, 66*(1), 66–75.

Gomà, R., & Rosetti, N. (1998). Ciutat Vella: Degradació, crisi i regeneració. *Barcelona i Societat, 9*, 108–139.

Lowndes, V., & Sullivan, H. (2008). How low can you go? Rationales and challenges for neighbourhood governance. *Public Administration, 86*(1), 53–74.

Lynch, K. (1960). *The image of the city*. Boston: MIT Press.

Marshall, T. (Ed.). (2004). *Transforming Barcelona*. London: Routledge.

Martí-Costa, M., & Cruz, H. (2009). La millora de la participació en els projectes integrals de la Llei de barris. In M. Martí-Costa & M. Parés (Eds.), *Llei de barris: Cap a una política de regeneració urbana participada i integral?* (pp. 137–146). Barcelona: Generalitat de Catalunya.

Martí-Costa, M., & Parés, M. (Eds.). (2009). *Llei de barris: Cap a una política de regeneració urbana participada i integral?* Barcelona: Generalitat de Catalunya.

Monclus, F. J. (2003). The Barcelona Model: an original formula? From 'Reconstruction' to strategic urban projects (1979–2004). *Planning Perspectives, 18*(4), 399–421.

Mongil, D. (2010). Intervención integral en barrios: Conceptos, instrumentos y elementos de mejora. *Ciudades, 13,* 139–161.

Montaner, J. M. (Ed.). (1999). *Barcelona 1979/2004: Del desarrollo a la ciudad de calidad.* Barcelona: Ajuntament de Barcelona.

Muñoz, F. (2006). Fer ciutat, construir territori: La llei de Barris. *Activitat Parlamentària, 8–9,* 59–75.

Nel·lo, O. (2008). Contra la segregación urbana y por la cohesion social: La ley de barrios de Catalunya. In Diputació de Barcelona (Ed.), *Ciutats en (re)construcció: Necessitats socials, transformació i millora de barris* (pp. 227–246). Barcelona: Diputació de Barcelona.

Nel·lo, O. (2009). *La llei de barris: Una aposta col·lectiva per la cohesió social.* Barcelona: Generalitat de Catalunya.

Nel·lo, O. (2010). The challenges of urban renewal: Ten lessons from the Catalan experience. *Análise Social, XLV*(197), 685–715.

Nel·lo, O. (2011). The five challenges of urban rehabilitation: The Catalan experience. *Urban Research & Practice, 4*(3), 308–325.

Nel·lo, O. (2012). *Ordenar el territorio: La experiencia de Barcelona y Cataluña.* Valencia: Tirant Humanidades.

Solà-Morales, M. (1987). La segunda historia del proyecto urbano: Otra tradición moderna. *UR: Urbanismo Revista, 5,* 21–27.

Vall, P. (2007). Matriu civica per a un corredor metropolità. In C. Mendoza & P. Vall (Eds.), *Granollers a tota escala* (pp. 17–21). Barcelona: ESARQ–UIC.

Vall, P., Koschinsky, J., & Mendoza, C. (2011). Retrofitting suburbia through pre-urban patterns: Introducing a European perspective. *UDI: Urban Design International, 16,* 171–187.

Chapter 5
The Port Resolution Project: Developing Community Built and Managed Visitor Accommodation Typologies

Damian Madigan and David Morris

Abstract This book chapter is a case study of a community development project undertaken by the University of South Australia's (UniSA) School of Art, Architecture and Design (AAD) for a village community at Port Resolution on the island of Tanna, Vanuatu. The project involves the design and construction of visitor accommodation, and continues an established program at UniSA's AAD school of engaging students in capacity-building development projects for remote communities. These have included projects for remote Aboriginal communities in central Australia. The Port Resolution project is the first design and construction project undertaken by AAD outside Australia, and the first to be designed as a typological exemplar that can be replicated by other communities on Tanna Island and potentially other islands across Vanuatu. Designed by students and to be built by students in association with the local villagers, the aim of these community run and owned facilities is to provide income to sponsor local education initiatives. The process described in this case study combines western and local architectural methods into a hybrid architectural practice. The chapter describes a design process wherein community engagement and immersion is foregrounded to become a much more significant component of a design process than may be anticipated at a project's inception.

Keywords Architecture · Cross-cultural design · Typological exemplar · Hybrid design

D. Madigan (✉) · D. Morris
School of Art, Architecture and Design, The University of South Australia, Adelaide, Australia
e-mail: Damian.Madigan@unisa.edu.au

D. Morris
e-mail: David.Morris@unisa.edu.au

R. Tiwari et al. (eds.), *M² Models and Methodologies for Community Engagement,* 69
DOI 10.1007/978-981-4585-11-8_5, © Springer Science+Business Media Singapore 2014

Introduction

The Port Resolution project is a community based visitor accommodation development which is an initiative aimed at generating income to further children's education for the Port Resolution community on Tanna Island, Vanuatu. The project is being assisted by staff and students from the University of South Australia's (UniSA) School of Art, Architecture and Design (AAD) and continues an established program of engaging students in capacity-building development projects for remote communities. It is hoped that this project will become an exemplar community-built and -managed enterprise, which can be replicated by other communities to generate income and to maintain local ownership and control of traditional land.

Vanuatu

Vanuatu is an archipelago of over 80 tropical islands in the South Pacific Ocean, located 1,750 km off the east coast of Australia and situated between the Equator and the Tropic of Capricorn. With a population of around 230,000 (Vanuatu National Statistics Office 2013), 80 % of the people live in rural areas, mostly as subsistence farmers in isolated villages (Asian Development Bank 2009). These geographically remote communities have minimal access to basic health and education services, and it is the cost of schooling in Vanuatu which is the central issue underlying the Port Resolution project.

Primary schooling in Vanuatu is free; however, secondary schooling requires annual fees of up to 60,000 Vatu per child (approximately AUD$660 at the time of writing), which is beyond the means of most families, limiting the opportunity for a child to complete secondary and therefore tertiary education. Churches and non-government organizations provide a minimal level of support to many rural villages. Such financial limitations perpetuate Vanuatu's dependence on aid and outside professional assistance.

Vanuatu's population, with an estimated growth rate of 2.3 % annually in 2009 (Vanuatu National Statistics Office 2013), is placing increased pressure on local resources, and this is coupled with impediments to economic growth such as geographic isolation and dispersion, vulnerability to natural disasters, poor infrastructure, limited export opportunities and a reliance on imports and tourism. Tourism accounts for around 20 % of GDP and 75 % of foreign exchange in the country (Asian Development Bank 2009). However, the inference that Vanuatu's limited economic opportunities have few positive consequences needs to be judged in the light of the measurable negative impacts from economic development, such as environmental degradation, and the cultural disruption caused by the incremental transfer of indigenous land ownership to commercial interests and capital investment from outside Vanuatu, clearly evident in the sale of land by agents in Port Vila.

Tanna Island

Tanna Island is the sixth largest island of the Vanuatu archipelago, with local estimates of population at about 23,000 people, spread mainly in the south and the west of the island, with far fewer settlements in the north and east. Tanna Island is one of Vanuatu's primary tourism destinations based around traditional culture and the active volcano at Mount Yasur to the east of the island. Mount Yasur is one of the world's most active and accessible volcanos. Further to the east of Mount Yasur is Port Resolution, named by Captain James Cook in 1774 after his ship *Resolution.* The port and surrounding coral reefs, secluded white sandy beaches, hot springs, lush tropical flora, and traditional villages attract fewer visitors due to the limited available accommodation. Visitors to Tanna Island commonly fly to the airport on the west coast and currently stay in resort-style lodges adjacent to the airport or the main town Lenakel, south along the coast. From there most visitors travel east over extremely rough four-wheel-drive tracks to see the Mount Yasur volcano, returning to the west coast on the same day.

The Port Resolution Project

The need for suitable visitor accommodation at Port Resolution was identified by the Chiefs of Tanna Island, in consultation with the local community land owners, as a sustainable business model for developing a trust fund for education by attracting greater visitor numbers to the east of Tanna Island. Given the constraints of travel from one side of the island to the other, marketing produce in Lenakel is unviable, however the tourism prospects of Port Resolution, because of its proximity to Mount Yasur, and the hot springs, white sand and coral reef, were agreed to be potentially more viable.

A high level of communication and cooperation between the chiefs and their communities across Tanna was important, since the financial benefit of such a proposal would initially be limited to the community at Port Resolution. This potentially divisive preferencing of one community over others needed to be understood in the context that the Port Resolution project would have a potentially broader benefit to the island as a whole, as an exemplar of community built and managed accommodation for other communities to replicate in the future.

The site chosen for the project is a beachfront to the east of the community of Port Resolution. The community is a village of approximately 300 people living in traditional, timber-framed thatched huts clad in panels woven from local pandanus and coconut leaves. Water is drawn from hand pumps connected to local wells and from streams in the surrounding mountains when available. Limited amounts of power are generated from solar panels and generators for pumping water and charging mobile phones. The majority of households grow their own food for subsistence and sell some at markets in the area. Limited income is derived from basic visitor accommodation and the sale of coffee, meals and artefacts.

Although a number of villages in the area provide overnight tourist accommodation, only a limited number of tourists visiting Mount Yasur choose to venture further east to stay overnight, due largely to the lack of accommodation options and a failure to meet visitor expectations. The identified need for visitor accommodation came to the attention of the University of South Australia's AAD Student Design and Construction Program, which has an established reputation for engaging students in capacity-building development projects for remote communities that might otherwise struggle to engage the services of an architect.

The Return Brief: Engagement and Community Consultation

The return brief is an evolutionary architectural tool that is vital to a project; it provides a means by which the designer can summarize their early understandings of the project requirements, constraints and possibilities in a feedback loop with their client. In essence, it is the opportunity to test one's interpretation of another's needs, expound the subtleties of the project as the architect understands them to be, and provide a vehicle by which the project's owners might confirm or reconsider their initial expectations. In the case of the Port Resolution project, the design team knew very little of its specifics before arriving on Tanna Island. However, this uncertainty—not knowing what to expect of the site, its custodians and the facilities they were asking us to design—was coupled with an open mind as to what might be possible.

What we did know was that the Port Resolution village community in particular and the Tanna Island community in general wanted a design for a community-run accommodation facility for tourists that would become a template for future developments across the island. Whilst very little of the specifics of the project were understood at its inception, what was clear was the importance of developing a relationship with the community as early as possible into our proposed 2 week working visit: the stronger the relationship, the easier it would be to develop a mutual understanding.

Such uncertainty and openness necessitates some relinquishing of control by the architectural team over a design project during the formative period; we had to be patient and resist the temptation to rush to a design solution in the absence of a nuanced understand of the project's physical and cultural contexts. For our students, this required the ability to question what appropriate architecture is when designing for communities other than their own, particularly where there is a clearly established vernacular building context. Often, prevailing western culture[1] can appear, at face value, to focus on building form and commodification as the drivers of architectural projects, and student designs can sometimes risk naïvely preferencing

[1] By 'western', we mean 'non-island' cultures generally, and more specifically, cultures such as Australia's, where disposable income is spent on tourism.

evocative forms and shapes over more subtle and restrained design approaches (in simple terms, a preferencing of form over function). In the case of the Port Resolution project, the architecture would be inescapably driven by the specific needs of a community and not the preferences of the architect. It was with this intention that we began working with a group of architecture and engineering students on potential generic strategies that might inform later specific decision-making once the full briefing process began on the island.

What resulted was a deliberate position regarding architectural 'fit'. As western visitors invited to contribute design expertise, we would approach the task with humility and empathy, reflecting what might be described as a hybrid practice: one where the prevailing architectural traditions of both cultures are allowed expression without being in competition. This hybrid process is described in the section 'The Design'.

The Visitor Experience

Whilst we understood and expected that a series of community consultation processes would occur, what was perhaps less anticipated was the extent to which a combination of formal and informal encounters with Tannese communities would inform the design. Upon arrival in Lenakel, it became apparent that the community wished first for us to understand what makes Tanna unique among Vanuatu's islands in order to appreciate not only why visitors might visit Tanna, but why the cultural heritage of the place—both tangible and intangible—was something to be shared with visitors. Importantly, it was this sharing of cultural heritage that would underpin the project; in continually describing village life, culture and history to visitors, one passes this cultural capital on to one's own future generations.

Thus began the visitor experience. Travelling first in the back of utility vehicles along challenging volcanic rock roads and then by foot through machete-cut tracks, the group was immediately immersed geographically and culturally in aspects of Tannese life by a visit to the Lekalangia *kastom* (traditional) village of Chief Yapa Nekiwris (Fig. 5.1).

Kastom life privileges traditional ways of living by restricting or forbidding the wearing of western clothes and avoiding contemporary technologies, but does not always exclude visits from outside parties. Chief Yapa's welcoming presentation to the group and generous tour and explanation of the village's private buildings were early and poignant markers that the Port Resolution project held the potential to bring an understanding of traditional Vanuatuan culture to a wider audience (Fig. 5.2).

Whilst deliberately resisting western influences, Lekalangia village welcomed us as foreign visitors as a way of celebrating and sustaining cultural difference. Importantly for the project, rather than diluting the tradition, deliberately attracting visitors through the design of a hybrid island/western accommodation model might bring about a combination of cultural identities that would strengthen both

Fig. 5.1 Walking through machete-cut tracks to the *kastom* village of Lekalangia: rarely visited by foreigners

Fig. 5.2 Part of the *kastom* village of Lekalangia

Fig. 5.3 An existing visitor accommodation bungalow on Tanna Island, showing a western-style ensuite off the sleeping quarters, divided only by a lightweight three-quarter height partition

heritages. In Chief Yapa's words to us, 'we might forge roads between Australia and Tanna.'

In a similar generosity of spirit, existing foreign-run accommodation facilities on the west of the island were made available for inspection. The owners were aware of the island-wide push to design alternative accommodation facilities that might provide self-sufficient income for villages; and despite being in direct competition with our project in the future, they were happy to share their knowledge and methods for running western-style tourist operations on Tanna (Figs. 5.3 and 5.4). These visits were crucial not only to determine the successes and failures of various project elements, but to provide evidence of areas where the Port Resolution project might offer points of difference. Continuing throughout the process, however, was a deep recognition that building a facility in itself is no guarantee of future business success, and gaining and retaining custom is an ongoing challenge.

Port Resolution

Fundamental to the island's ability to attract visitors in the first instance is Mount Yasur; the difficult trek to which is rewarded with panoramic views over Tanna's east coast before a descent through dense bush and then arrival on the vast ash plains at the volcano's base (Fig. 5.5). In a somewhat surreal experience of driving

Fig. 5.4 A Tanna Island 'honeymoon suite' containing an ensuite bathroom visible from the sleeping quarters

Fig. 5.5 After the difficult trek to Mount Yasur, seen on the right with ash plume, one is rewarded with panoramic views of Tanna's east coast (Port Resolution is beyond the ridges behind the volcano)

through an ash desert along the sides of the volcano, the visitor is brought into proximity with the island's most prominent physical and spiritual entity. Significantly, it is this journey, in immediate contact with the volcano, that one takes to reach Port Resolution.

Our arrival in the community was marked by two unexpected and powerfully affecting events to thank us for making the journey to assist with the project: the singing of welcoming songs and the provision of a meal in a purpose-built temporary pavilion. These simple reciprocal acts of greeting and thanks immediately established that both parties to the project were appreciative of the opportunity to be engaged with each other and the process that was to come (Fig. 5.6). As visitors, this community-led welcome placed us in the position of future patrons of the new visitor accommodation facilities, and this visitor experience was to continue over the first few days, as we were taken to the volcano's rim, the hot springs it produces (which are regularly used for cooking) and around the bay in which Captain Cook

Fig. 5.6 Simple reciprocal acts of arrival and thanks immediately established that both parties to the project were appreciative of the opportunity to be engaged with each other

landed and walked in 1774. It became apparent during this process that taking the time to understand Port Resolution's unique characteristics was a crucial part of the community's briefing process, as important to the community as the eventual formal consultation process to follow. Integral to this social manoeuvre was time spent in the *nakamal*.

The *Nakamal*

Vanuatuan villages are arranged around an open circular kava-drinking ground called the *nakamal*, usually an area established beneath a large banyan tree. Every evening around dusk the men of the village meet in the *nakamal* to prepare the kava, an infusion of the root of *piper methysticum*. In this ritualistic process, the men clean the roots of loose dirt using coconut husks, before chewing the root and infusing the resultant pulp with water to form a drink. This is then taken whilst making silent offerings to the spirits for wellbeing. This is a male-only activity, and we were advised that the men of our group might be invited to the *nakamal* for kava, and that this act would be the single biggest gesture of good will by both parties. The invitation occurred on our second night in the village and continued every night thereafter.

In addition to communicating with the spirits, the kava preparation and drinking ritual sees the *nakamal* become a place of daily reflection and communication on matters of governance and social and spiritual order (Fig. 5.7).

This occurs in softly spoken dialogues that eventually become whispers and then silence before the men gradually leave the *nakamal*. In an architectural sense, this daily event, for which all other work in the village ceases, provided us with an opportunity to reflect on the work in progress, both internally as a design team and with our clients. What for our clients is a fundamental means by which to run village life was for us the opportunity to critically engage with the work of the day in the absence of distraction and to seek feedback and direction from our clients. Something akin to a formal design review process in Australia, the sunset withdrawal to the *nakamal* was an obligatory yet welcome period of project reflection detached from the physical processes of architectural production.

For the women of our design team, the men's kava time became an opportunity to engage with the women of the community in private, learning traditional skills and establishing relationships with the village's children through play. This demarcation of male and female activities, whilst unexpected as a daily occurrence, enabled the design team to regroup each evening to share what we had learnt from our hosts (Fig. 5.8).

Importantly, especially during the early phases of our time on Tanna, the combined acts of hospitality and inclusion gave us opportunities to be immersed in village life in a meaningful way, as well as a deeper understanding of the project's requirements and potential before we were tempted to rush to a design solution. Of the short two weeks we had to engage with the community and design, and to

Fig. 5.7 Learning to clean kava roots of dirt with coconut husks in the *nakamal*; a place of daily reflection and communication on matters of governance and social and spiritual order

Fig. 5.8 The design team regrouping to work during the night

Fig. 5.9 The first formal briefing with the community, which occurred after a series of less tangible project briefing activities

develop and deliver a concept to them on site, the first week passed without the need for a formal drawing to be produced. And whilst such initial immersion in the culture of the place might be seen as a reduction in valuable production time, the opposite proved to be true. Indeed, rushing to design drawings in the absence of this immersion would risk the design being prematurely and naively developed.

The Design

> …[T]radition is a complex continuity inherited from the past, lived in the present and sustained in the future (Oliver 2006, p. 384).

Combined with the less tangible project-briefing activities of experiencing village life, a formal community briefing was held whereby an open discussion could be had around issues of day-to-day running of the facility, accommodation requirements, meal preparation, laundering, building servicing and maintenance (Fig. 5.9).

This was immediately followed by a physical survey of the site, including the location of all trees, existing features and site contours. It was perhaps this surveying activity that for the first time introduced the community to formal design thinking and the skills the design team would bring to the project. Where in the past buildings in the village have appeared in an ad-hoc manner as needs dictate, this was an

Fig. 5.10 The design team's surveying activity introduced to the village for the first time the role of formal design thinking

exercise in gathering data in order to record the site conditions and strategically master plan an appropriate built response at the scale of a small village (Fig. 5.10).

This formal master planning exercise was vital, as upon completion, the project will encompass:

- three types of bungalow accommodation (six bungalows in total);
- a main communal building consisting of restaurant, bar, kitchen, office, welcoming area and ablution facilities;
- an outdoor kitchen accommodating traditional village cooking methods;
- photovoltaic and wind power generation;
- water supply and waste water management systems; and
- interpretive signage within the immediate accommodation site and beyond (Fig. 5.11).

Less tangible, yet significant elements are embedded design approaches to cross ventilation, privacy, views, hygiene, traditional cooking requirements and vernacular and western building technologies.

Fig. 5.11 The Port Resolution visitor accommodation facility showing the proposed scheme of bungalows and main hall, with existing site features, levels, trees, grass, sand and water positions recorded during the site survey

In considering the site layout and the resultant buildings that would form the visitor facility, the design intent was to produce something that might be described as 'similar but different' to the vernacular Tannese buildings. Returning to the concept of the architectural hybrid, designing in a village context requires an avoidance of romanticizing or historicizing the existing architecture in favour of a deliberate effort to meld two cultural approaches. The architectural result is, therefore, something that is not in essence from either culture in itself, but is a coming together of the two.

This raises the question of why one would not design a traditional Vanuatuan building. For us the answer lies in two facts. The first is that the hybrid approach is already the norm when applied to other visitor accommodation on the island, most noticeable in bungalows of masonry construction (as opposed to being traditionally timber framed and clad) and with western-style ensuite bathrooms (as opposed to detached communal sanitary facilities of lightweight construction). The second factor is that traditional buildings are best done by the local community without

Fig. 5.12 A masonry ensuite bathroom in an existing visitor accommodation bungalow on Tanna Island provides the convenience of proximity, but has problems of overall amenity and privacy

external input. The reality of the Port Resolution project is that external expertise was sought specifically for the differences that that process can deliver. The resulting design, therefore, is one that seeks to augment the traditional islander building methods with imported western architectural understanding at the scales of the master plan, the building, the room and the construction detail.

This design-led value-adding, important as it is, given the fact that visitor accommodation on the island is already being provided in a manner that attempts to preempt what western visitors will expect, is evidenced in various ways throughout the Port Resolution design scheme. In the bungalows (visitor accommodation), ensuites are provided as they are in other visitor accommodation on Tanna, but in a manner that provides greater amenity and privacy, whilst considering the implications for the initial construction requirements, regular cleaning regime and ongoing maintenance and repair (Figs. 5.12 and 5.13).

In the main hall, traditional outdoor fire-pit cooking methods can be maintained, but in a safer, more comfortable and more convenient manner for the village women, who will inevitably run the kitchen (Figs. 5.14 and 5.15).

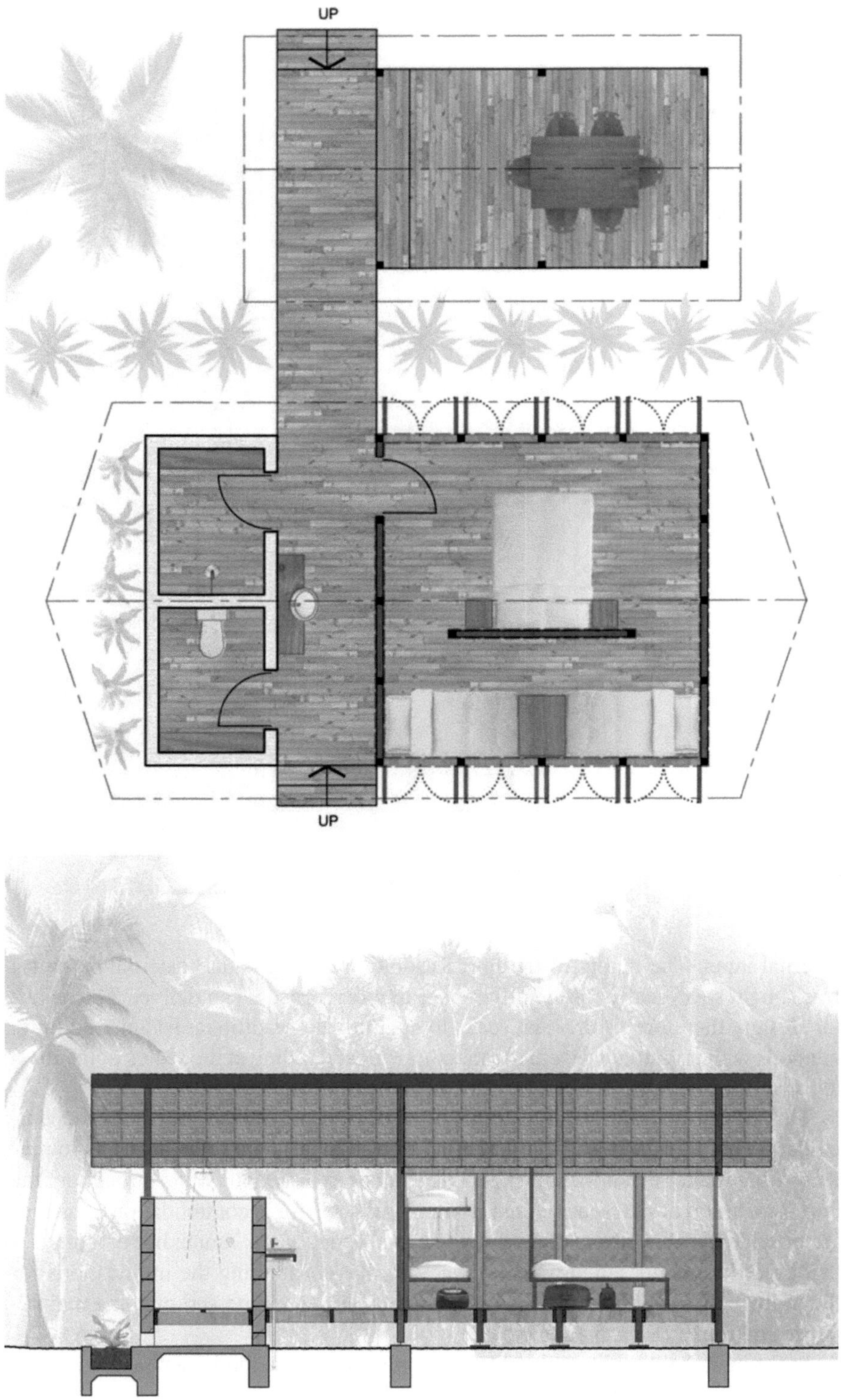

Fig. 5.13 Similar but different: one of the three proposed bungalow types in plan (*above*) and cross section (*below*) showing sleeping quarters separated from ablutions via the use of a breezeway deck: the convenience of proximity is maintained, but privacy, amenity, sanitation, and ease of maintenance are significantly improved

Fig. 5.14 Traditional cooking methods at the project site, where women cook in the open over a fire grate in a ditch whilst squatting down

Fig. 5.15 Simlar but different: a physical model of the outdoor kitchen with the roof removed to reveal and explain how the interior will work: logs are continuously fed in large lenghts immediately into the fire from an adjacent store, minimising wood cutting, and benches are provided for working at more comfortably

On a site scale, an analysis of power generation requirements and wastewater management systems aims to service the visitor accommodation efficiently and reliably, whilst contributing to the living standards of the village as a whole.

Such whole-of-village design thinking links the project to its broader ambitions of a sustainability strategy that operates architecturally, environmentally, economically and culturally.

Architectural Communication

Successfully communicating architectural design intentions requires the translation of design thinking with clarity. Orthographic drawings such as floor plans, external wall elevations and building sections often require a significant level of visual interpretation in order to be successfully read, even by those familiar with the genre. However, in western cultures, floor plans at the very least are a part of everyday life and are seen in real estate advertising, wayfinding devices in shopping centres and in evacuation plans in public buildings. Even outside the discipline of architecture, the floorplan is, for many, a readable and understandable communication tool for converting two-dimensional information into visualised three-dimensional space. For the Port Resolution project, where the facilities' owners were unlikely to have ever seen a floorplan before, a two-stage presentation method was used: one for on the island and another for wider distribution upon our return to Australia.

Hand drawings produced on site were used internally by the design team to explore and communicate ideas to each other, and whilst these were presented to the community formally before departure as a means of describing our work process and as an introduction to the way architectural drawings are used to think about and explain buildings, it was a physical model that was used to describe the scheme (Fig. 5.16).

Working with a combination of model making materials such as cardboard and balsa wood brought with us from Australia, and supplementary materials found on site, such as twigs to represent trees and pandanus leaves for thatched rooves, the design team modelled the site terrain, trees, grasses, sand and water, existing structures and the proposed new buildings. Roofs of the proposed buildings could be removed to reveal the spaces within. This physical tool provided a tangible mechanism by which the village community could grasp what was being proposed; seeing existing grass, trees and paths in their actual locations, albeit in miniature, made clear the concepts being put on offer. Whilst the drawings were brought back to Australia to enable further design work to be undertaken, the model was left with the village for their use as a discussion tool amongst the community and as a communication tool to show to visitors.

Upon return to Adelaide, the design team undertook a three-week design workshop in which the preliminary concepts established on site were developed using a suite of two-dimensional drawings, three-dimensional computer renderings, and a detailed physical model. These more resolved drawings were then used in concert

Fig. 5.16 Here, the roof of the modelled main hall is removed whilst the rest of the model is used to explain what the spaces in the hall will be, and how they will operate

with the model to bring the project to a wider audience in both Australia and Vanuatu, for the promotion and preliminary fund-raising for the project (Figs. 5.17 and 5.18). It is not enough to simply design a scheme and present it evocatively; proposals such as this require confidence building amongst those who have the potential to help fund and deliver the project.

Future Directions

At the time of writing, the design proposal was being developed in detail by staff and students from academic programs in architecture, interior architecture, industrial design and civil engineering, with the objective of determining a detailed resolution of the design that will decide costs for future funding applications and enable the construction of the project on site.

The areas of the design requiring particular resolution include:

- architectural detailing of storage and kitchen facilities, wet areas, fittings, fixtures, finishes, furniture and hardware;
- structural engineering of building frames to resist cyclonic winds;
- water supply pumping, storage and reticulation;
- wastewater treatment and reuse; and
- power generation, storage, inversion and reticulation.

Fig. 5.17 The developed concept design exhibited at the School of Art, Architecture and Design at the University of South Australia

Fig. 5.18 A photorendering of the proposed dining room with existing beach and ocean visible through the windows to the right

The design development strategy focuses on achieving a built outcome which is both environmentally and culturally sustainable, by maximising available materials and building skills, and minimising the costs and reliance on imported equipment and technologies. Embedded in the design is an understanding that the facilities are to be constructed using a combination of skilled and unskilled labour, and that ongoing maintenance and upkeep needs to be inexpensive and manageable for the local community, whilst minimizing or avoiding reliance on external support. The design strategy, which recognizes that many visitor accommodation facilities in Vanuatu are foreign-owned, is intended to be an exemplar for other communities seeking financial independence through local community enterprise, and as a means of retaining indigenous land ownership against potential capital investment from outside Vanuatu.

In the future, it is hoped that staff and students will be involved in assisting the Port Resolution community to construct the project, as part of AAD's Student Design and Construction Program. However, the initial design of this project can only establish its spatial and material order—it cannot, by itself, ensure its successful delivery. Ultimately the venture is being established and operated by a community inexperienced in this type of development, and it is therefore anticipated that the project will require further work with the community to build capacity to organize, manage, and maintain the Port Resolution project, without relying on outside assistance. In this regard, it is anticipated that broader disciplines within UniSA will become part of an expanded project team, particularly in regard to health, marketing, tourism and business management.

Capacity Building

Fundamental to the notion of capacity building is an appreciation of existing capacity. Within the Port Resolution community, this was evident during our first visit, where 16 staff and students were very comfortably accommodated for 10 days in bungalows with beds and clean sheets, and catered for with three meals a day of local food such as yams, taro, beans, sweet potato, spinach, chicken, pork and fish, and an equally vast choice of fruit such as bananas, coconut, melon, and paw paw. However, in order to achieve construction and servicing standards to meet visitor preferences for up to 30 overnight visitors and possibly 50 conference attendees, the capacity to service visitors on such a scale underlies the purpose of the cross-cultural engagement between the community and the University to address particular capacities, including:

- **organizational capacity**: the management of marketing, staff rosters, visitor bookings, visitor arrivals, tours, supplies, cooking, laundering, cleaning and waste management and disposal;
- **financial capacity**: the management of money including receipt of payments, banking, budgets, payment of bills and salaries and developing a system of equitable benefit sharing based on effort, underpinned by a strategy of saving to cushion against fluctuations in visitor demand and unexpected costs;
- **health and hygiene**: the development of cleaning regimes and food preparation protocols which build upon established practices, but which become increasingly important in maintaining the health and expectations of numerous visitors day after day;

- **technical capacity**: the development of existing timber construction and masonry techniques to achieve high levels of structural integrity and serviceability, and the development of expertise to repair and maintain power systems, refrigerators, generators, pumps, plumbing, waste water treatment systems and buildings.

Further to these aspirations, the project has already—in the first iteration of what will most likely be a 5–10 year process—developed capacities:

- **within the community**: through the ability to witness and understand formal planning processes, design and spatial conceptualisation and strategic thinking around greater project implications; and
- **amongst our students**: through placing them in the roles of professional practitioners working with real clients on a tangible project, across cultures and in the field, away from their familiar office-based design tools.

The Port Resolution project arose out of a collective aspiration of the Tannese people to develop a community-managed tourism enterprise to achieve a sustainable business model for developing a trust fund for education. If the hybrid design strategy and the capacity of the community combined make the project a success, it will have achieved its aim as a model for other community managed enterprises wishing to provide greater educational opportunities to their children. The ultimate result of this merging of professional architectural processes with existing cultural knowledge and sensitivities will, therefore, be less dependence on foreign professional assistance in the future, and the preservation of the community's ownership of their land, with minimal impact on what is a fragile tropical paradise.

Acknowledgements For his assistance and guidance on the project in general and for his background information on traditional Tannese life, we gratefully acknowledge the assistance of Mark Nizette, PhD candidate in Pacific and Asian History at the Australian National University.

References

Asian Development Bank. (2009). Vanuatu Economic Report 2009: Accelerating Reform. http://www.adb.org/publications/vanuatu-economic-report-2009-accelerating-reform. Accessed 17 June 2013.

Vanuatu National Statistics Office. (2013). Social Summary. http://www.vnso.gov.vu/index.php/social-statistics. Accessed 17 June 2013.

Oliver, P. (2006). *Built to meet needs: Cultural issues in vernacular architecture*. Burlington: Architectural Press.

Chapter 6
Applying a Practical, Participatory Action Research Framework for Producing Knowledge, Action and Change in Communities: A Health Case Study from Gujarat, Western India

Clancy Read, Jaya Earnest, Mohammed Ali and Veena Poonacha

Abstract Participatory action research (PAR) has an extensive history in many fields of social practice. The applied research methodology is change focused, and works towards improving a problem that has originated in the community itself; the problem is then further defined, analysed and solved by participants. PAR promotes mutually respectful relationships, shared responsibilities, and an emphasis on local capacity building that promotes environments in which communities increase their ability to uncover local barriers and harness local assets to build healthier communities.

This chapter presents a modified PAR framework that can effectively guide participatory action researchers from multiple disciplines working with communities to produce knowledge, action and change, whilst empowering communities to construct and use their own knowledge to emancipate themselves from their situations—the goal of PAR.

Being context-specific, participatory approaches to research can be applied to any study with its own methodology. A case study undertaken in Western India will exemplify how this practical PAR framework can be applied to assess and

C. Read (✉) · J. Earnest · M. Ali
Centre for International Health, Curtin University, Perth, Australia
e-mail: clancyread@hotmail.com

J. Earnest
e-mail: j.earnest@curtin.edu.au

M. Ali
e-mail: M.Ali@curtin.edu.au

V. Poonacha
Research Center for Women's Studies, Dr. Avabai Wadia Archives, SNDT Women's University, Mumbai, India
e-mail: veena_poonacha@yahoo.co.uk

R. Tiwari et al. (eds.), *M² Models and Methodologies for Community Engagement*,
DOI 10.1007/978-981-4585-11-8_6, © Springer Science+Business Media Singapore 2014

address health problems in disadvantaged communities. The study setting and context presented is in a rural district of the southern Gujarat State in India, composed predominantly of tribal communities.

Keywords Participatory action research · Health needs assessment · Intervention · Community-based · Change · Rural health

Introduction

Participatory action research (PAR) and its 'sister concept, participatory research (PR) involves social investigation, education and action in a collaborative and interrelated process' (Hall 2005, p. 5). Researchers and participants collectively undertake self-reflective inquiry so they can 'create new forms of knowledge through a creative synthesis of the different understandings and experiences of those who take part' (Rice and Ezzy 1999, p. 173), and understand and take action to improve the practices in which they participate and the situations in which they find themselves. 'The [applied] reflective process is directly linked to action, influenced by understanding of history, culture, and local context and embedded in social relationships' (Baum et al. 2006, p. 854).

The applied methodological framework presented in this chapter is a practical approach to working with communities to improve a problem that has originated in the community itself, and is to be defined, analysed and solved by participants. Modifying the traditional PAR model of planning, acting, observing and reflecting, researchers are guided through a multistage, context-specific approach for working with communities, from knowledge generation to action and—as a result—change. The framework can be applied by researchers from multiple disciplines whose research places communities and their participation as a central role in the research process.

The modified PAR framework promotes mutually respectful relationships and shared responsibilities, whilst emphasizing local capacity building, and empowering communities to construct and use their own knowledge to emancipate themselves from their situations—the goal of PAR. This proposed modified framework brings new improvements to practice, and its combination of theory and practical considerations makes it suitable for field based practitioners and academics alike. Following this brief introduction, the chapter gives an overview of the study setting and objectives.

The Study Setting and Overarching Objective

Research collaboration was established in 2008 between the fourth author, Director of the Center for Rural Development (CRD) in the Research Center for Women's Studies (RCWS) at SNDT Women's University, Mumbai, India, and the second author, from the Centre for International Health (CIH) in the Faculty of Health Sciences, Curtin University, Western Australia.

The initial requirement of CRD was an assessment to gain a better understanding of available health care facilities and access to the health of local communities in the southernmost district of Gujarat State in Western India.

A research proposal was jointly conceptualized and funding was received for the research from the Centre for Advanced Studies in Asia, Australia and the Pacific in the Faculty of Humanities in Curtin University. The first author later conceptualized, designed and conducted the research as part of her PhD studies and was supervised by the second and third authors. Using a multi-stage approach, the objective of the study was to explore community perceptions of health issues in five rural, predominantly tribal villages in Gujarat, Western India, and transform community knowledge into action with the meaningful participation of community members.

The *Taluka* (administrative sub-division) where the research was conducted comprises 79 villages and three towns. Thirty-nine percent of the population of 405,902 belong to Scheduled Tribes (2001 Census, Table 17). According to the latest recorded estimates (2002) the area had a total Below Poverty Line (BPL) population of 16,149 families and the majority of these families were Scheduled Tribes (Government of Gujarat n. d.). Scheduled Tribes are recognized in the constitution of India as a group of historically disadvantaged people who require affirmative government policies and programs for their development (Government of India, Ministry of Health 2001).

Thinking Practically: Modifying the PAR Framework

Based on reflection and review of research throughout the duration of the first author's PhD study, the conceptual framework applied in this study was modified and further developed. The result is presented diagrammatically in Fig. 6.1.

The modified PAR framework was used for this study. The remainder of this chapter will describe each stage of the applied framework and the participatory approaches to research that were adopted.

The Framework Structure and Guiding Methodology Explained

In the published literature, the ideology of PAR is often discussed, rather than a description of what actually happens in the field (Rice and Ezzy 1999, p. 178). PAR is a context-specific methodology, so participatory approaches to research with their own set of methods can be chosen and applied to the specific context of the study. It is possible to describe what actually happens during the participatory approaches overlaid into this framework rather than PAR itself. This PAR framework guides the researcher through the process of knowledge and action generation, fostering constant reflection throughout the process.

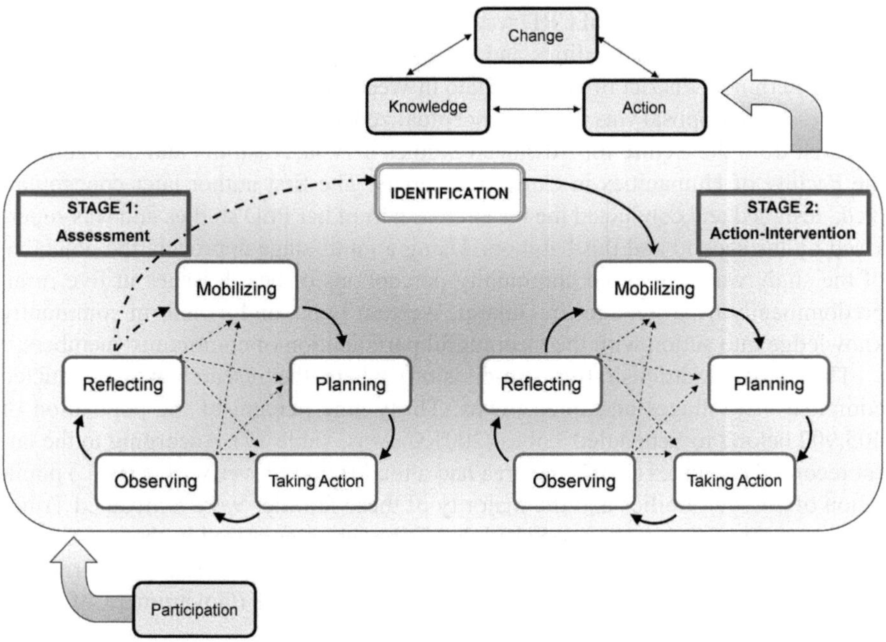

Fig. 6.1 From assessment to action-Intervention: a practical approach to PAR (Read 2012, p. 284)

According to the principles for implementing PAR in community research, 'the problem originates in the community/organization itself and is defined, analysed, and solved by the participants' (Balcazar et al. 2004, pp. 22–24). This reinforces the two main stages of the PAR framework: identifying the problem and addressing it with action.

Amongst some of the benefits of this methodology is that the deliberative nature of PAR enables researchers and communities to understand and transform practices in which they participate, as well as the situations in which they find themselves, through critical and self-critical action and reflection. It should be noted that in recognition of the iterative nature of PAR, the entire process outlined in Fig. 6.1 is intended to be repeated as many times as it takes to achieve the desired results, as mutually defined and agreed upon by the organization and community. Further cycles will allow for evaluation of changes.

The 'Ins and Outs' of the Practical Framework

Whilst there are many expected inputs of PAR, a study without participation is not PAR; thus participation is a mandatory input to the process. PAR may start with small groups of collaborators but widens so that it gradually includes more and more of those involved and affected by the practices in question. Researcher-community partnership renders results more accessible, accountable, and relevant to people's lives (Israel et al. 1998). Through active engagement, individuals and communities may become more empowered and better equipped to make sustainable personal and social change (Wallerstein and Duran 2003).

Ideally, the goal of the completed cycles of PAR is change; however, change is not the only expected output of this proposed practical approach to PAR. The modified PAR framework represents the relationship between knowledge, action and change. All three elements can be an outcome of the entire PAR process; however, the main expected outcome of Stage 1 (Fig. 6.1) is to generate knowledge through the application of a chosen participatory approach.

With this knowledge as a basis, the participatory approach to research applied in Stage 2 (Fig. 6.1) would work to transform this knowledge into action. Action followed by change would be a specific expected output of Stage 2, which is action-intervention undertaken to address the community problems. Change would again inform knowledge and action. The relationship between the three elements flows in both directions and is reciprocal. Change informs action, which informs knowledge, continuing the cyclic relationship between the three elements.

Apart from knowledge, action and change, the process can generate valuable information, awareness and partnerships. Increased capacity and development of the researchers and communities can result from the learning process. Other benefits may include joint reports, possibility for further funding, capacity building of in-country researchers, awareness of limitations in the developing world, and a chance for communities to share their stories and voices.

Rewards or benefits to the community can be expected from this PAR. This benefit is balanced with the gains to the researcher/organization and should also be balanced with the level of risk involved in participating in the study. Joint ownership of the study as well as empowerment of communities should be fostered. Trust can be earned from communities and the process can build awareness amongst communities in addition to knowledge, action and change.

Commencing the Study/research: Stage 1, Assessment

In order to commence action to address an issue or situation or improve the well-being of communities, a topic must first be identified that may be initiated by the community and/or the organization/researcher/practitioner. It is recommended that the assessment stage be undertaken regardless of how the topic is identified. This stage provides an opportunity to further explore the complexities of the topic at hand, determine if the topic is a priority for the community, and promote understanding in preparation for Stage 2. This stage involves considerable exploration of the community-identified problem and dialogue as to whether the community wishes to work towards improving and resolving the problem.

The commencement point of the first PAR self-reflective cycle is mobilizing. Without first mobilizing research partners, participants and resources, the study or programme cannot commence. Once mobilization has been achieved, the study progresses to the planning and action phases. Observation and reflection occur at all phases, as represented by the arrows linking all phases of the cycles. In reality, after mobilizing, the activity at each phase of the cycle may not necessarily occur in order as the diagram suggests, and it is possible the phases may overlap and occur simultaneously.

Another point to note is the dotted arrows from the reflecting phase, with one returning to mobilizing and the other progressing to identification, showing the link between the two stages. This is to remind researchers/practitioners that assessment is a fluid stage and its repetition can be beneficial to continual knowledge generation. The dotted arrow to identification indicates that quality information was generated from the assessment stage, with which action-interventions can be initiated.

Application of Stage 1: A 'Real World' Example

A rapid participatory appraisal (RPA) was conducted to fulfil the requirements of Stage 1, Health Needs Assessment in Gujarat, Western India. This participatory approach to research was deemed appropriate, as it is employed as a means of assessing needs prior to the preparation of plans for future action (Annett and Rifkin 1995, p. 6). It is the first step in the planning process, thus suitable for the commencement of a PAR study.

Throughout the process, the World Health Organization's (2002) Guidelines for Rapid Participatory Appraisals to Assess Community Health Needs (cited in Annett and Rifkin 1995) were consulted. Eight stages of the RPA were applied to the modified version of Denzin and Lincoln's cyclic model of PAR. The resulting framework can be viewed in Fig. 6.2.

The health information pyramid proposed in Annett & Rifkin's (1995) guidelines was adopted as a guide and framework to determine what information would need to be collected in the appraisal process (Fig. 6.3).

The information pyramid consists of four levels, and nine blocks of information that are used to assess community health needs. Information was built from the bottom up; information about community composition, organization and capacities to act formed the foundation of the appraisal. This provided information about the strengths and weaknesses of the community leadership, organizations and structures in the villages. The second level of information was concerned with developing a profile of those aspects of the community's environment that have major implications for health. The third level of the information pyramid concerned data on the existence, coverage, accessibility and acceptability of services. The final layer detailed national, regional, and local policies that indicate the level of government or political commitment to community participation in health. Information from each level and block of the information pyramid helped develop the assessment of the health status and identification of health problems in the selected villages in Pardi Taluka and determine priorities for health development and services.

Information was collected from each village on the nine health-related aspects of the information pyramid, using multiple methods for rapid analysis and community participation: semi-structured interviews with key informants and focus groups

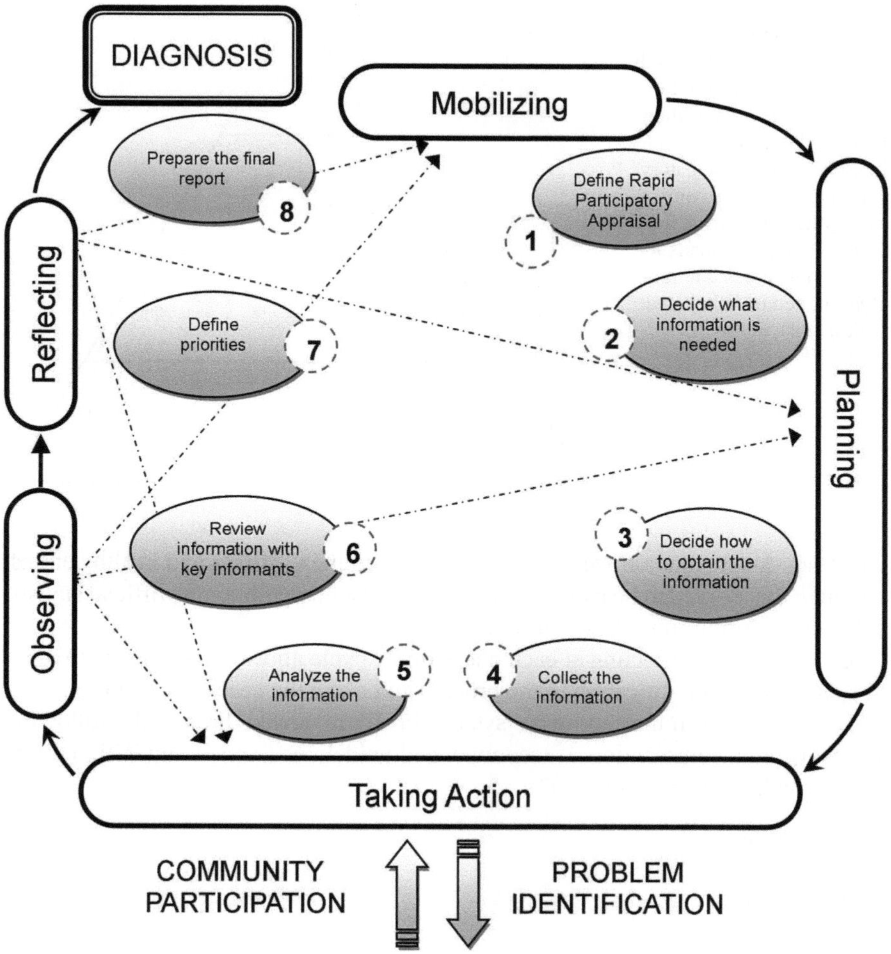

Fig. 6.2 Steps of rapid participatory appraisal in a PAR study (Read 2012, p. 104)

drawn from the local community; direct observation through community visits; and data from secondary sources. In total, 82 key informants and focus group participants were interviewed throughout the RPA. The major source of information was obtained from interviews with key informants who were selected because they were in the best position to understand the main health issues of the community. These informants were selected by the team in the field, and included: local government officials, teachers, social and health service personnel, and community leaders.

To gain a deeper understanding of the community's priority health and social needs, particularly from the perspective of village and tribal women, focus groups were conducted. Women from pre-existing self-help groups in the community were selected to be involved in these focus groups. At least one focus group took place in each village, with up to 12 women in each group. Existing written records, reports

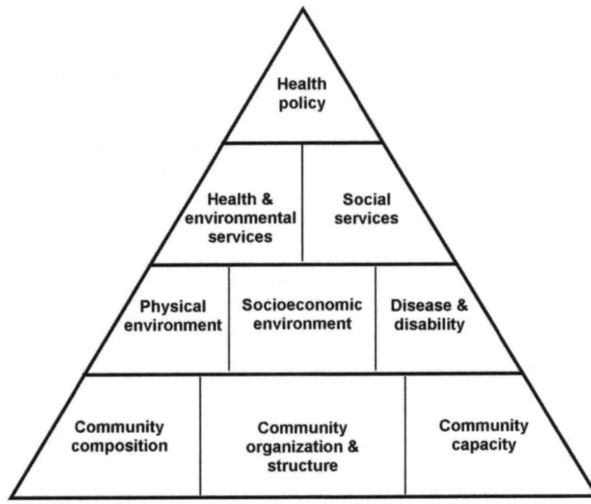

Fig. 6.3 Blocks for an information pyramid for use in assessing community health and social needs; from the World Health Organization published document: Guidelines for rapid participatory appraisals to assess community health needs: a focus on health improvements for low-income urban and rural areas. (Cited in Annett and Rifkin 1995, p. 10)

and other documentation were sought from government offices and health centres, although this proved to be a difficult exercise, particularly obtaining official statistical data in writing from these sources.

Observing and reflecting at each phase of the cycle allowed researchers to identify the need for modification to the process and make the necessary changes.

Data retrieved in the field were systematically reviewed through the following stages, using a combination of recommended guidelines (Annett and Rifkin 1995; World Health Organization 2002):

- Comparing
- Collating
- Reviewing
- Summarizing
- Interpreting

First, information from different sources was compared and checked for discrepancies. Next, data was collated under categorized themes and subjected to thematic content analysis. Once the information was agreed by researchers to be a correct reflection of the information collected, data was summarized into concise statements. It was then possible to interpret the findings and list the main problems for each village. Documented problems were recorded as stated by the community, not as interpreted by the researchers.

Given the time constraints of the rapid assessment process, this analysis was undertaken by the researchers and later confirmed and prioritized with participants in the identification phase of the framework, described in the following section. In this study, the RPA approach was a viable method for assessing community health needs in rural, indigenous, resource-poor settings in a developing country. Detailed knowledge about health, environmental and socio-economic issues was generated, providing a holistic approach to health needs assessment and a foundation of information on which to commence Stage 2, Action-Intervention.

Fig. 6.4 Listening to the local knowledge of women in one of the villages in Pardi Taluka, Gujarat State, India

Fig. 6.5 A community meeting led by the village Sarpanch (leader). On occasions when village elders were in attendance, women would take a lesser role in the meeting activities

Bridging the Gap Between Knowledge and Action: Identification

The process of confirming and prioritizing results of Stage 1 with communities is an important step in validating results. It serves as an effective link between the two main stages of the study (Stage 1: Assessment and Stage 2: Action-Intervention) and its importance in the knowledge to action process should be represented accordingly. Identification, shown in Fig. 6.1, occurs as its own step between stages and cycles. This step is given its own label to remind researchers to take time to return to participants of Stage 1, to confirm and prioritize results and share results with the communities. The results of this process will influence Stage 2 of the study. Dialogue generated during this process is necessary to explore problems and can often result in generating ideas about what action could be taken to address the problems. This process validates the results researchers have analysed, and ensures that identified issues are locally defined and prioritized. Involving communities in this process is beneficial, as it supports collaboration and allows insights into how communities perceive their local problems.

Application of 'Identification': Confirming and Prioritizing

Upon completion of the analysis phase of the RPA, Annett and Rifkin (1995) advise researchers to return to the key informants to convey results and ask their opinion about the priority they give to different problems that have been defined by researchers during the analysis phase.

Accordingly, in the applied study, each key informant was presented with the list of problems for their village. They were asked to prioritize each of the problems with a number, starting with one, until all problems were numbered. They were also asked to note if any actions were being undertaken to improve the corresponding problem, and to add any further comments they had. These two additional questions revealed if action was already underway to improve the problems, and gave opportunity for key informants to comment on their village's problems.

This prioritizing exercise was extremely important. If researchers had incorrectly identified a problem in the village, then the ranking exercise would result in a low priority being allocated to that problem by respondents. The result of this confirming step was a succinct list of problems for each village, prioritized by community members themselves. With this information researchers could then commence planning for Stage 2, Action-Intervention.

Stage 2: Action-Intervention

For the same reasons as Stage 1, Stage 2 commences with mobilizing. It is then a continual process of mobilizing, planning and taking action that is accompanied by reflection and observation, and is intended to be constantly repeated, incorporating what was learned during each cycle into the next one. In this stage of the PAR, action is a major expected output of the process, with action followed by change.

Application of Stage 2: Addressing Identified Problems with Action

In this study, a community based participatory research (CBPR) approach was chosen as the most appropriate one to be applied to the PAR model to implement action. CBPR provides a structure and mechanism for collaborative and rigorous research, using well-established or emerging methods with a community focus (Horowitz et al. 2009). In public health, research focuses on social, structural, and physical environmental inequalities. Through active involvement of, for example, community members, organizational representatives, and researchers in all aspects of the research process, all partners contribute expertise and share decision-making and responsibilities (Israel et al. 2005, p. 1464; 1998; p. 173; 2001, p. 182; 2003, p. 54). 'Partners contribute their expertise to enhance understanding of a given phenomenon and integrate the knowledge gained with action' to enhance the health and wellbeing of community members (Israel et al. 2001, p. 182).

Applying this participatory approach, the aim of Stage 2 was to transform knowledge of community problems identified in Stage 1 into action, as suggested by the communities in their participation. Observing and reflecting at the mobilizing phase of Stage 2 allowed researchers to analyse the process of the PAR and guide research accordingly.

To mobilize communities, village leaders were consulted to start with. In those villages where village leaders were supportive of pursuing action to address identified problems, the participatory method of community consultations was organized. These consultations were designed to allow community members to guide the progression of the research with support from researchers; however, throughout the process, numerous challenges to the PAR process emerged. These challenges were documented, and the feasibility of conducting PAR in the villages was assessed. Due to a number of complex interrelated factors and the nature of the challenges, to be discussed in the following section, we could not achieve mobilization of community members, and the PAR cycle of Stage 2, action-intervention did not eventuate.

Lessons Learned for Future Practice

This section draws on participant views and field experiences to share lessons learned with practitioners. The reflective process made clear that what eventuated from this study is the result of a myriad of interrelated contextual factors impacting the transformation of knowledge to action (Fig. 6.6).

A number of factors categorized as environmental, community, organizational, and practitioner/researcher issues affected the transfer of knowledge into action, despite the soundness of the participatory approach applied to the PAR cycle.

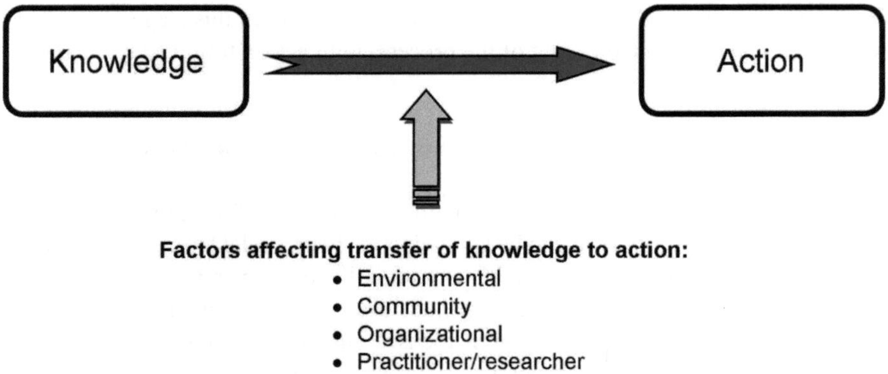

Fig. 6.6 The knowledge to action process and factors affecting its transfer (Read 2012)

Environmental factors relate to the situation in which the research is being applied. In this study, the risk in participating in action-interventions to address identified problems was present, and was constantly discussed by participants and highlighted by the in-country research assistant. The communities also had negative past experiences of taking action on the problem being addressed, which affected their motivation to participate. It is a recognized challenge of CBPR that 'participation in the action phase of CBPR projects may sometimes present risks to community participants and actions that involve challenging powerful corporate or other entrenched interests may have negative consequences for those involved' (Minkler 2005, p. ii10). Risk has multiple meanings; it can be actual physical harm that will result from a participant's or community's participation; it can also be the repercussions of not being able to fulfil daily duties because the person is investing time in participating in an action-intervention. Both factors were present for participants in this study. The risks were intensified by incidences of corruption, intimidation and violence during past activities which had sought to address long standing community problems.

Organizational factors can also affect the ability to transfer knowledge to action. These included paucity of resources and difficulties facing the organization/initiators of change in carrying out action-intervention. Another barrier is a low level of trust between the community and the organization/initiator of change. In this study, there were budgetary and human resources constraints in working with communities at the intensity required of the PAR style, together with the nature of problems revealed in the RPA. Whilst the collaborating organization had links with communities, their work in the study villages was mainly with small groups of women. The organization had not previously worked with villages in their entirety, so rapport with other stakeholders within the community had not yet been established, and this contributed to the challenges in translating knowledge to action.

Community participation is a central component of PAR. Several sub-factors related to the community can hinder capacity of action-interventions, especially in achieving participation. In this study, the community's only engagement with the organization was from the RPA conducted in Stage 1 of the research. A sustained longer-term relationship with the community could have built trust, ensured engagement and possibly resulted in community efforts. Researchers sensed that the problems uncovered in the needs assessment of Stage 1, whilst important enough to reveal and discuss, were not a priority for the key informants, and this may have contributed to their reluctance to participate in the interventions. Personal matters, such as earning an income to feed the family, attending to seasonal crop work, and social commitments including weddings, illness, and festivals, took priority over community issues. Time too is critical and must be factored into studies for events such as these, as should the season at the time of the study.

Motivation to participate in activity was further affected, due to the lack of consensus about the problem being addressed. Many participants saw the problems as personal rather than of community concern, and thus they were not motivated to work on them. These issues, together with the lack of support from village leaders, and the newness of a participatory approach to solving community problems, were among the challenges we faced in transferring knowledge to action in this study.

Factors related to the practitioner and researcher can hinder a participatory action-intervention. These relate to the researchers themselves, and the effects of power imbalance between researcher and participant, as well as the time allocated for the project. 'Even outsiders who pride themselves on being community allies and trusted friends frequently fail to realize the extent of the power imbued by their own, often multiple, sources of privilege and how it can adversely affect interactions and outcomes' (Fadem et al. 2003, p. 254). The time-consuming nature of CBPR is continually mentioned in the literature, and is viewed as 'one simple manifestation of insider-outsider tensions that may involve conducting a participatory community project within the timeframe of the academic and funding calendar' (Fadem et al. 2003, p. 254).

Real collaboration takes a lot of time: for meetings, for accountability processes, for working through the inevitable conflicts (Stoeker 2003, p. 101), and to establish and maintain trusting relationships (Israel et al. 1998, p. 182). In this study, the time required to mobilize community members, amongst other factors, was severely underestimated. The primary researcher was on an academic timeframe, and a scholarship that is only funded for a specific duration; this is not conducive to the nature of participatory research.

Each factor described, categorized as the research environment, organization, community, and researcher, played an important role in the outcome of this research. The impact they had presents an important lesson for researchers engaged in participatory action and community-based participatory research.

Concluding Thoughts

This chapter has presented a practical approach to applying PAR in communities, from assessment of community needs, to an example of addressing the identified needs in a case study conducted in five villages in Southern Gujarat in Western

India. When applied in a health context, rapid participatory appraisal and CBPR are appropriate participatory approaches that can be applied to and used with the modified PAR model. The identification stage of the model was an extremely important process to ensure results reflected community perceptions, and empowered communities to prioritize their needs and made them reflect on possible solutions. Knowledge to action transition can however be impeded by external and contextual factors that must be considered when conducting and implementing an action oriented research project. This study makes a call for further research so that these contextual factors can be better understood, and researchers can be prepared for potential contextual challenges prior to commencing the action-intervention stage of the modified PAR framework.

Acknowledgements This study was undertaken as a collaborative community project with the Centre for Rural Development (CRD) and SNDT Women's University in Mumbai, India, in collaboration with the Centre for International Health (CIH) at Curtin University, Western Australia.

Funding for this study was provided by the Centre for Advanced Studies in Asia, Australia and the Pacific (CASAAP) and in-kind support was provided by the CRD.

The primary researcher expresses gratitude to all employees at the Center for Rural Development, Udwada who assisted with this project. A special acknowledgement is made to Suman Bali, past Coordinator, Health Project, whose support during the research was invaluable.

This study would not have been possible without the participation of the communities of the five villages in Southern Gujarat, and we sincerely thank those who participated and gave their time.

References

Annett, H., & Rifkin, S. (1995). Guidelines for rapid participatory appraisals to assess community health needs: A focus on health improvements for low-income urban and rural areas. Geneva: World Health Organization, Division of Strengthening of Health Services. http://whqlibdoc.who.int/hq/1995/WHO_SHS_DHS_95.8.pdf. Accessed 8 Aug 2008.

Balcazar, F. E., Taylor, R. R., Kielhofner, G. W., Tamley, K., Benziger, T., Carlin, N., & Johnson, S. (2004). Participatory action research: General principles and a study with a chronic health condition. In L. Jason, C. Keys, Y. Suarez-Balcazar, R. R. Taylor & M. I. Davis (Eds.), *Participatory community research: Theories and methods in action* (pp. 17–35). Washington, DC: American Psychological Association.

Baum, F., MacDougall, C., & Smith, D. (2006). Participatory action research. *Journal of Epidemiology and Community Health, 60*(10), 854. doi:10.1136/jech.2004.028662.

Fadem, P., Minkler, M., Perry, M., Blum, K., Moore, L., & Rogers, J. (2003). Ethical challenges in community based participatory research. In M. Minkler & N. Wallerstein (Eds.), *Community based participatory research for health* (pp. 242–262). San Francisco: Jossey-Bass.

Government of Gujarat. (n. d.). Taluka at a Glance. http://www.vanbandhukalyanyojana.gujarat.gov.in/taluka%20profileFinal/Valsad/Pardi.pdf. Accessed 9 Aug 2011.

Government of India: Ministry of Health. (2001). Census of India 2001: Scheduled Castes & Scheduled Tribes Population, 2001, Office of the Registrar General and Census Commissioner.: Delhi. http://censusindia.gov.in/Census_Data_2001/India_at_glance/scst.aspx. Accessed 9 Sept 2011.

Hall, B. (2005). In from the cold? Reflections on participatory research from 1970–2005. *Convergence, 38*(1), 5–24. http://www.proquest.com. Accessed 9 Aug 2011.

Horowitz, C. R., Robinson, M., & Seifer, S. (2009). Community-based participatory research: From the margin to the mainstream. *Circulation, 119*(19), 2633–2642. doi:10.1161/†CIRCULATIONAHA.107.729863.

Israel, B., Parker, E. A., Rowe, Z., Salvatore, A., Minkler, M., Lopez, J., & Halstead, S., et al. (2005). Community-based participatory research: Lessons learned from the Centers for Children's Environmental Health and Disease Prevention Research. *Environmental Health Perspectives, 113*(10), 1463–1471. doi:10.1289/ehp.7675.

Israel, B., Schulz, A. J., Parker, E. A., & Becker, A. B. (1998). Review of community-based research: Assessing partnership approaches to improve public health. *Annual Review of Public Health, 19,* 173–202. doi:10.1146/annurev.publhealth.19.1.173.

Israel, B., Schulz, A. J., Parker, E. A., & Becker, A. B. (2001). Community-based participatory research: Policy recommendations for promoting a partnership approach in health research. *Education for Health, 14*(2), 182–197. doi:10.1080/13576280110051055.

Israel, B., Schulz, A. J., Parker, E. A., Becker, A. B., Allen III, A. J., & Guzman, J. R. (2003). Critical issues in developing and following community based participatory research principles. In M. Minkler & N. Wallerstein (Eds.), *Community based participatory research for health* (pp. 53–76). San Francisco: Jossey-Bass.

Minkler, M. (2005). Community based research partnerships: Challenges and opportunities. *Journal of Urban Health: Bulletin of the New York Academy of Medicine, 82*(2), ii3–ii12. doi:10.1093/jurban/jti034.

Read, C. (2012). Applying a participatory action research model to assess and address community health concerns among tribal communities in Gujarat, Western India: The potential and challenges of participatory approaches (Doctoral thesis, Curtin University). Retrieved from Australian Digital Thesis Website. http://espace.library.curtin.edu.au:80/R?func=dbin-jump-full&local_base=gn01-era02&object_id=188776. Accessed 21 April 2013.

Rice, P. L., & Ezzy, D. (1999). *Qualitative research methods: A health focus*. Melbourne: Oxford University Press.

Stoeker, R. (2003). Are academics irrelevant? Approaches and roles for scholars in community-based participatory research. In M. Minkler & N. Wallerstein (Eds.), *Community based participatory research for health* (pp. 98–112). San Francisco: Jossey-Bass.

Wallerstein, N., & Duran, B. (2003). The conceptual, historical, and practice roots of community based participatory research and related participatory traditions. In M. Minkler & N. Wallerstein (Eds.), *Community Based Participatory Research for Health* (pp. 27–52). San Francisco: Jossey-Bass.

World Health Organization. (2002). *Community participation in local health and sustainable development: Approaches and techniques. European Sustainable Development and Health Series*. World Health Organization Regional Office for Europe.

Chapter 7
Inspired by Nature: Building Community Capacity Through Creative Leadership

Denise K. DeLuca

Abstract Building the capacity of communities to achieve sustainable design requires developing community leaders who can choreograph diverse teams of creative, idealistic people able to realize radical innovation in a complex and rapidly changing world. A pilot online course called Creative Leadership was taught with the goal of empowering sustainable design students in a Master's degree program to become effective leaders.

Sustainable design requires radical innovation. Conventional leadership is very good at driving incremental innovation, but is ill suited to drive and deliver radical innovation. Conventional leadership is also inadequate in increasingly interconnected, complex, unpredictable, and volatile conditions.

Creative Leadership uses nature as a model for leadership principles and practices, because evolution has yielded organisms that are responsive, adaptable, resilient, able to leverage and support dynamic complex systems, and embrace unpredictability and disturbance. A Creative Leader continually maintains and refreshes the common vision, values, and goals; sensing and adapting to changing conditions both inside and outside the community; and creating conditions of trust, respect, and curiosity that enable dynamic synergisms and co-creativity.

The course covered: classical, emerging, tribal, and nature's leadership models, organizational structures, and cultural stages; methods for leading creative people and projects; understanding oneself and others; and envisioning oneself as a Creative Leader. The course also included methodologies leaders can use to facilitate positive, co-creative, emergent discussions, as well as to develop a culture of trust, respect, and curiosity, including: reflection in nature; creative visualization; feedback metrics; 'yes, and'; attentive listening; Socratic and collaborative inquiry; and creating triad-based networks. Initial feedback from students suggests a shift in thinking and increased capacity to lead in creative, idealistic, and challenging situations.

D. K. DeLuca (✉)
BCI: Biomimicry for Creative Innovation, Seattle, USA
e-mail: denise@businessinspiredbynature.com

R. Tiwari et al. (eds.), *M² Models and Methodologies for Community Engagement*,
DOI 10.1007/978-981-4585-11-8_7,

Keywords Leadership · Creative Leadership · Sustainable Design · Radical innovation · Natural Leadership · Leadership Inspired by Nature · BCI · Biomimicry

Pilot Project: Creative Leadership Course

The communities targeted in this chapter are those engaged in driving sustainability or implementing sustainable design solutions. This chapter is based on a pilot online course called Creative Leadership, taught as part of the Master of Arts degree in Sustainable Design Online (SDO) offered at the Minneapolis College of Art and Design (MCAD). The goal of the course was to empower emerging sustainable designers to become effective leaders and, in doing so, to build the capacity of the communities with which they work to achieve sustainability goals.

Sustainability inherently requires the integrated efforts of diverse communities, as it must simultaneously address the interconnected yet sometimes conflicting issues of business, nature, and humanity. Successfully achieving sustainability goals—especially those that go beyond simply doing less bad—requires these diverse communities to have clear common vision, values, and goals, and the ability to work together synergistically and co-creatively. A Creative Leader builds the capacity of these communities by continually maintaining and refreshing the common vision, values, and goals; sensing and adapting to changing conditions both inside and outside of the community; and creating conditions of trust, respect, and curiosity that enable dynamic synergisms and co-creativity.

This chapter will discuss: the need for Creative Leadership; the method for developing sustainability leaders through this program and course; the concept of Leadership Inspired by Nature, which was the model for this course; selected methodologies taught in the course; and the nascent capacity building outcomes of the pilot course.

The Need for Creative Leadership

Achieving radical innovation requires leadership skills and approaches that differ from conventional leadership (Slater et al. 2013). As Warren Bennis puts it, 'One person can live on a desert island without leadership. Two people, if they are totally compatible, could probably get along and even progress. If there are three or more, someone has to take the lead' (Bennis 2009, p. 4). As the world becomes more interconnected, complex, unpredictable, and volatile, we need strong leaders more than ever (Bennis 2009); however, conventional leaders find themselves struggling to run their communities and organizations amid this growing tumult, and are beginning to realize that traditional organizational structures and leadership models are no longer sufficient (Chakravarthy and Henderson 2007; Safian 2012).

Conventional leadership models, born of the military and honed in the industrial age, are designed for efficiency and productivity (Doyle and Smith 2001; Safian

2012). Conventional leadership (leader as lone hero) and conventional organizational structures (top-down hierarchies) are based on the assumptions that an organization is controllable and that one person can be in control, yet organizations operate as part of a world of complex systems that are inherently uncontrollable (Wheatley 2011).

Wheatley (2011) argues that we need to foster leaders that act as hosts rather than heroes, that understand complexity and systems, and that invite participation and contribution by all members. 'Hosting leaders create substantive change by relying on everyone's creativity, commitment and generosity' (Wheatley 2011, p. 16).

Sustainable design represents a combination of ideals and ideas; however, people who are driven by ideas and ideals tend to struggle with the pragmatics of completing and implementing projects (Belsky 2010). The speed, scope, and scale of change needed to achieve sustainability require radical rather than incremental innovation (DeLuca 2012c). Conventional leadership is very good at driving incremental innovation, but is ill suited to drive and deliver radical innovation (Christensen 1997).

New types of leaders, able to succeed in a rapidly changing world, are beginning to emerge. 'Flux leaders' thrive in an environment of change and chaos, embrace 'adaptability and flexibility; an openness to learning from anywhere; decisiveness tempered by the knowledge that business life today can shift radically every 3 months or so…' (Safian 2012, n.p.). Although these flux leaders appear to exhibit characteristics needed to navigate our new paradigm, case studies of these new leadership models are few and are not yet time-tested.

Nature has extremely high quality control standards—only 0.1 % of species that have ever existed have survived the radical changes and disruptions that have occurred on Earth over the past few billion years. This suggests that organisms that exist today reflect time-tested strategies for surviving change and disruption (Benyus 1997). Evolution has yielded organisms that are responsive, adaptable, resilient, able to leverage and support dynamic, complex systems, and embrace unpredictability and disturbance (Pascale et al. 2000). These are the qualities that are exhibited by today's emerging successful leaders (Safian 2012) and needed to lead communities that are seeking sustainable solutions (DeLuca 2012b).

The premise of this chapter is that building the capacity of communities to achieve sustainability requires developing community leaders that can choreograph diverse teams of creative, idealistic people able to realize radical innovation in a complex and rapidly changing world. The Creative Leadership course uses Nature as a model for the leadership principles and practices needed to build the capacity of communities to forward and realize sustainable design solutions.

Method for Building Capacity

MCAD launched the MA in Sustainable Design in 2012 as part of the Sustainable Design Online (SDO) program. Program faculty recognized that graduates tend to lack the leadership skills needed to forward sustainability ideas and ideals, thus they chose to include a leadership course in the program. Cindy Gilbert, Director of MCAD's SDO Program, recognized that an alternative to conventional leadership was needed and, together with the author, envisioned a 10-week course based on Biomimicry for Creative Innovation's (BCI) concept of Leadership Inspired by Nature (DeLuca 2012b).

The first half of the course focused on developing knowledge and understanding of different leadership models, organizational structures, and cultural stages (from classical and emerging to tribal and inspired by nature), and methods for leading creative people and projects. The second half focused on students learning more about themselves, understanding others, and envisioning themselves as Creative Leaders. The course also developed skills needed for Creative Leadership, a selection of which is described in the methodologies section.

The Model: Leadership Inspired by Nature

Leadership Inspired by Nature was developed to empower leaders to transform their organizations to become in harmony with Nature, as well as to leverage the benefits of reflecting the principles and practices of Nature (DeLuca 2012b). In this section, we present Nature's Principles and leadership models found in Nature.

Nature's Principles

Leaders inspired by Nature seek to embed and reflect Nature's Principles in the organizations and communities that they lead. Nature's Principles, adapted from the design-focused Life's Principles (Benyus 1997), are: resilient, optimizing, adaptable, systems-based, values-based, and life-supporting (Biomimicry for Creative Innovation 2011) (see Fig. 7.1)

Being resilient is the ability to recover from a disturbance and to leverage disturbance as an opportunity, rather than trying to protect against change. Resilience is achieved by creating a diverse, dynamic, and co-creative culture. **Optimizing,** in contrast to the conventional minimizing or maximizing, requires recognizing whole dynamic systems rather than focusing on single metrics or individuals. **Adaptability** requires being curious, continually sensing and responding to change, being flexible, creative, and continually moving towards positive outcomes. **Systems-based** means recognizing that every individual, organization, product, and process is part of numerous dynamic interdependent systems. Leaders inspired by nature foster synergies and build integrated value-generating networks. **Values-based** means having, and always acting in accordance with, a vision and a clear set of values. Leaders inspired by nature are the keepers of an organization's vision and values, and use them to align the energy and passion of its members, and to create a culture of mutual respect and trust. **Life-supporting** means working as an active and beneficial part of nature and humanity. Leaders inspired by nature recognize needs and goals beyond themselves, simple profit, or sustainability.

Fig. 7.1 Nature's Principles

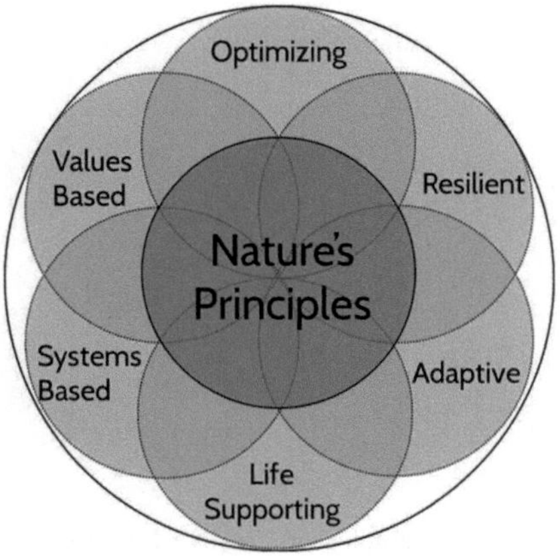

A Leader Inspired by Nature seeks synergies rather than independence; fosters co-creativity rather than competition; leverages collective intelligence rather than forcing top-down decisions; strives to be open-source rather than closed-source; optimizes across the many rather than maximizing for the few; seeks return on engagement rather than return on investment; thinks and works in dynamic systems rather than in static linear chains; creates organizations that fit form to function rather than expecting people to function within a rigid pre-set form; considers the very long term while addressing the very short term; favours emergence over sticking to pre-set plans; and seeks to be effective, not just efficient (DeLuca 2012a).

Methods for Creative Leadership

Practising Creative Leadership requires different skills than expressed by the conventional leaders who tend to be students' role models. This section presents selected methodologies used in the course to develop skills that emerging leaders need to choreograph diverse teams of creative, idealistic people who are able to realize radical innovation in a complex and rapidly changing world. These methodologies can be applied to facilitate positive, co-creative, emergent discussions, as well as to develop a culture of trust, respect, and curiosity.

Reflection

Conventional leaders are often valued for their inherent/apparent intuition, independence, charisma, power, and ability to spontaneously act and react. By contrast, Creative Leaders are valued for their empathy, collaboration, openness, drive to empower their networks, and ability to sense and respond thoughtfully to the dynamics of their systems. These characteristics can be enhanced by routinely devoting time to reflect on their vision and values, themselves, others, and the dynamics of

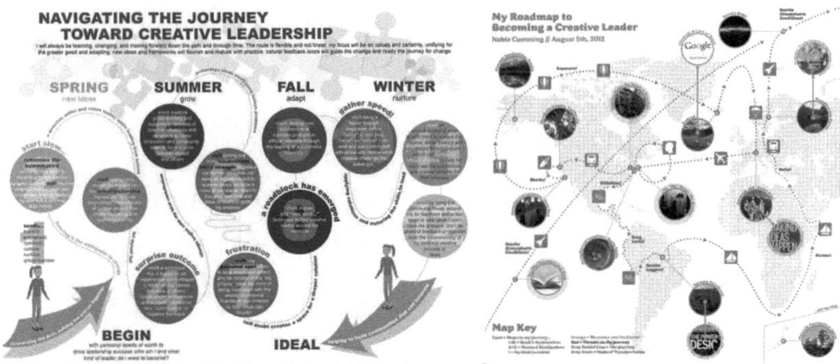

a b

Fig. 7.2 Examples of Creative Leadership journey 'maps' created by students. **a** by Shannon Rahkola. **b** by Noble Cumming

systems in which they work, including Nature. In class, students learn and practice three different forms of reflection: reflection in nature, keeping a 'travel log', and creative visualization.

Reflection in nature involves spending 30 min in nature: the first 10 min just 'being' (sitting, doing nothing) with eyes closed and mind open; the second 10 min observing (still just sitting, but now using all senses to just notice what is going on around them) and contemplating a pre-selected topic with eyes and mind open; and the last 10 min imagining ideas, solutions, outcomes, and so on, related to the pre-selected topic. This activity helps calm a chaotic mind, awaken dull senses, and re-connect with Nature and our inner Nature in order to have more thoughtful, creative, and expansive sessions through contemplating and imagining.

The ***travel log*** is a series of weekly journal entries, kept in a personal website, that students use to chronicle their journey to becoming a creative leader, often written immediately after their reflection in nature. At the end of the course, students use these entries to create a 'map' that illustrates their journey through the course as well as where they want to go in the future. Figure 7.2 contains images from the personalized maps created by two of the students in the class. Each map contains real or metaphorical locations, with descriptions of why and how visiting these places can help them be or become a better creative leader. Routinely articulating reflections and progress helps students clarify and embed what they have learned. Students are encouraged to read each other's travel logs to gain different perspectives on weekly lessons and personal revelations.

Creative visualization involves actively imagining being and/or expressing desired leadership characteristics in a given situation. This practice allows students to safely practice (in their minds) the application of desired characteristics in different and perhaps challenging settings and scenarios. They can use the results of creative visualization to envision or adjust how they might realize or express desired characteristics, and to gain confidence as leaders.

Measuring What You Value, Setting Up Effective Personal Feedback Loops

People tend to value what they measure, so it is important that we measure what we value. Conventional leaders tend to value the easily measured metrics of money and time and ignore externalities or things that are not measured. Creative Leaders continually assess how well they and their organizations are adhering to and reflecting their core vision and values (which are far more complex and comprehensive than just money and time) in real time and over time in response to changing conditions and context. In order to do this, leaders must have a way to measure their values, characteristics, behaviour, and/or performance in real time and/or over time as conditions change.

Students in the course are required to generate metrics (quantitative measurements) for a set of leadership characteristics that are most important to their personal development as leaders. Each metric includes a description, a unit of measure, and an appropriate measuring device. These metrics are intended to give them real-time and/or long-term feedback on leadership qualities they value and want to develop.

For example, a student might decide 'being a good listener' is the desired characteristic. The description might be 'quality as listener'; the unit of measure might be 'percentage of my responses that are in the form of a question rather than a statement' (thus indicating I was really listening rather than simply waiting for my turn to talk); and the measuring device might be a weekly bar graph, with % questioning on the Y-axis and daily conversations on the X-axis.

They use their set of metrics to create a Creative Leadership 'dashboard'. For class, the devices and dashboards were designed, but not constructed. The students were then asked to provide snapshots of dashboard readings they might imagine under various leadership scenarios. Figure 7.3 includes snapshots of dashboards created by two different students. Each contains a set of measuring devices along with descriptions of the characteristics they are measuring and how they interpret and respond to the readings.

This activity develops leadership capacity in several ways. First, students must identify a short list of leadership characteristics that they most want to personally develop. Then they must come up with ways to quantitatively measure these qualitative characteristics. This is a struggle for most, and forces students to engage much more deeply with both the meaning and the effective expression of their values. Finally, they have to reflect on and envision how they, as creative leaders, might actually respond and express their values in a range of scenarios.

'Yes, and'

Conventional leaders tend to rely on top-down, one-way communication and directives. Conventional group discussions tend to take the form of problem identification and solving, and offense/defense. Even the practice of brainstorming tends to be dominated by the ideas of the few and/or leads to consensus ideas.

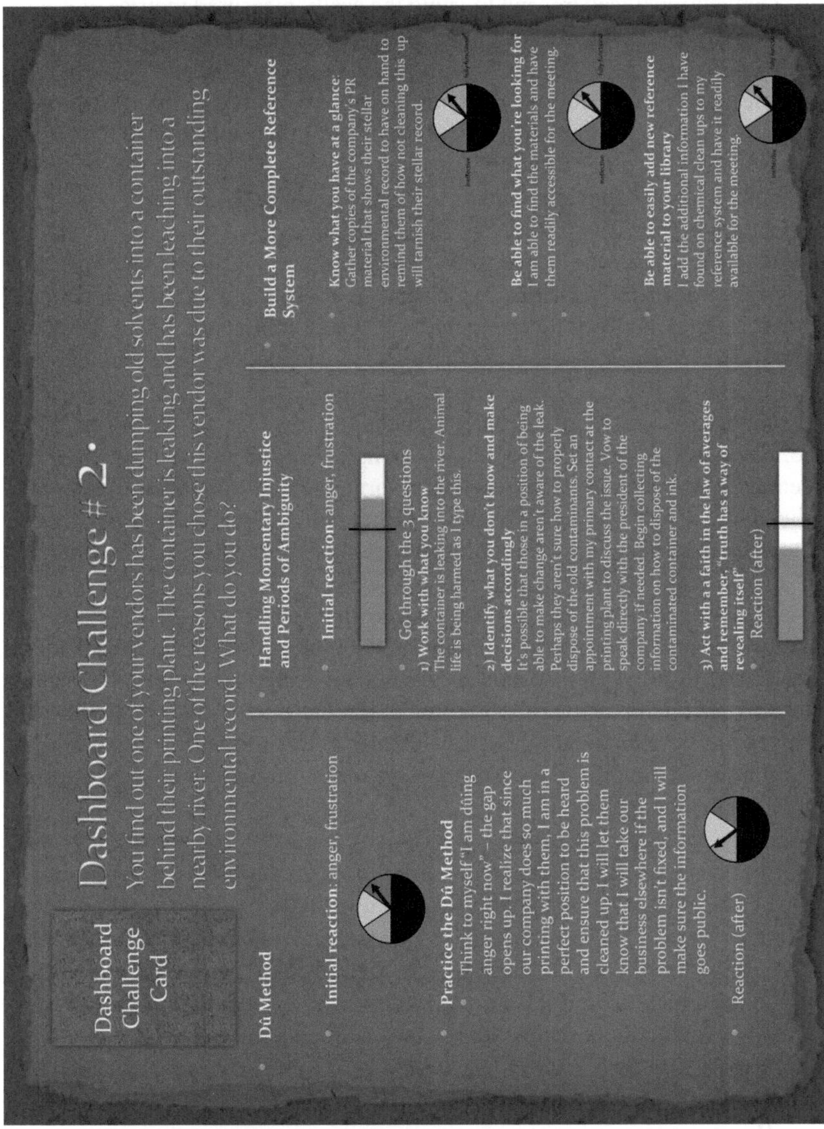

Fig. 7.3 Examples of personal Creative Leadership 'dashboards' created by students. **a** by Shari Welsh. **b** by Noble Cumming

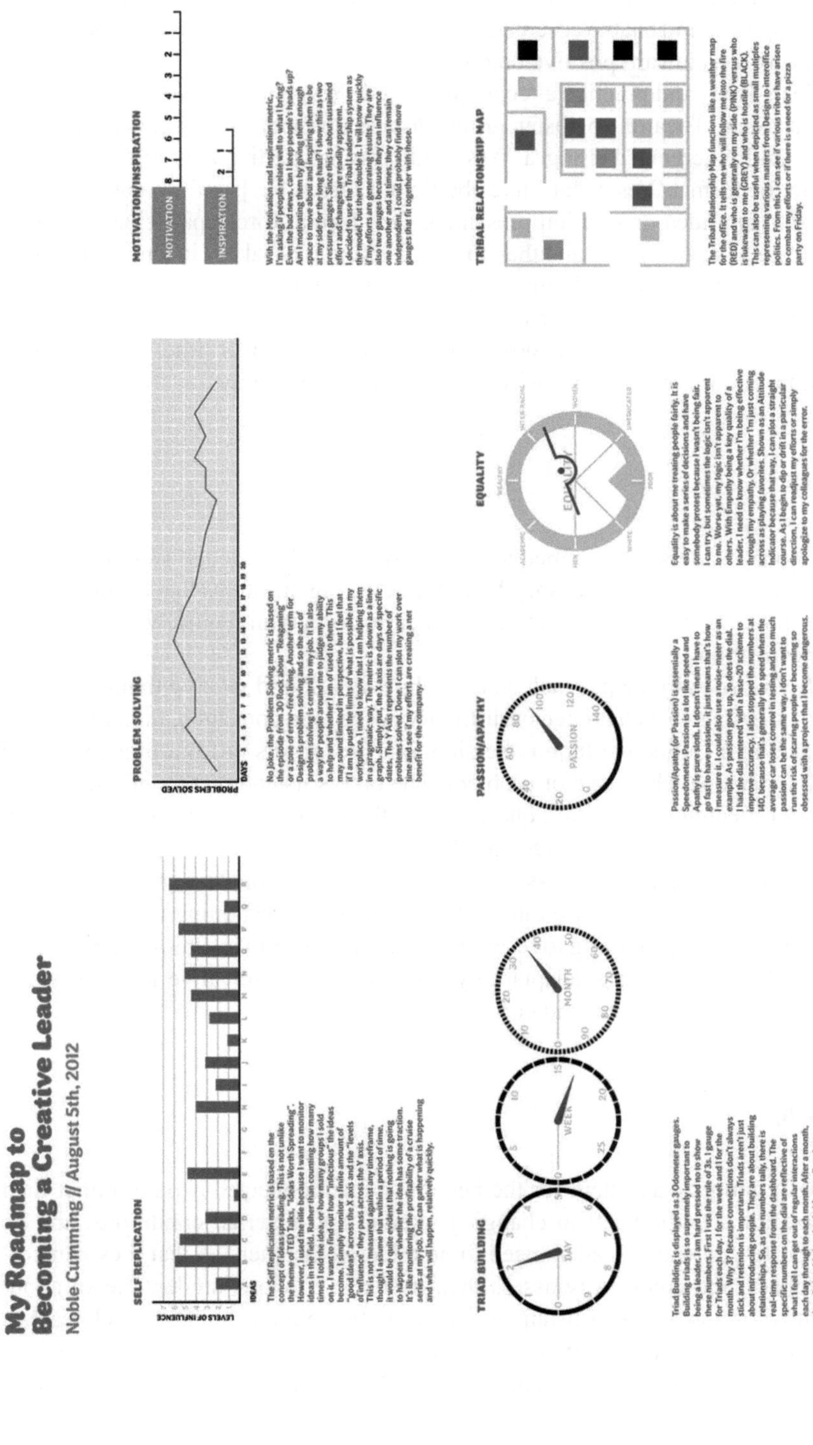

Fig. 7.3 (continued)

Sustainable design requires radical, emergent solutions co-created among a diversity of thinkers, some of whom may have conflicting or fixed ideas. Creative leaders empower and enable co-creativity within their organizations and continually drive discussions toward positive outcomes.

Students learn and practice a technique called 'Yes, and', derived from Applied Improvisational Theatre, to foster group discussions that are creative, inclusive, positive, and emergent. During a 'Yes, and' session, participants in a discussion adhere to the format 'Yes, what I like about your idea is [...], and we could [...]' in response to any idea or solution presented. This format forces participants to first re-articulate a previous idea and then to either add to it or take it in a new direction.

In the short term, this technique is a tremendous tool for catalysing productive and co-creative discussions, even among people who otherwise tend not to get along together or tend not participate. When routinely practised over time, this technique can build a culture of trust, curiosity, respect, and co-creativity.

Inquiry

Conventional leaders are assumed to be the holders of knowledge and thus tend to give answers. Creative leaders spend more time asking questions that encourage expansive thinking, reflection, creativity, and curiosity within themselves and within their organization.

Collective Inquiry helps teach students how to ask and receive thoughtful empathetic questions to a challenge or difficulty posed by a colleague, rather than giving advice. In a Collective Inquiry session, one person sits in front of a group and presents a personal challenge or problem. The others listen quietly and then spend 5–10 min reflecting and writing questions for the presenter. All questions are then read out aloud while the presenter simply listens, without answering or defending.

During a ***Socratic Inquiry*** session, students learn to facilitate a group discussion where everyone is asked to use three different lines of questioning to explore a selected issue, rather than trying to resolve or find an answer to it (Phillips 2001). This develops skills in listening, adaptability, and emergent co-creative thinking while building a culture of respect and curiosity.

Listening

In order to optimize a system for the benefit of interconnected people (rather than maximize for a few), adapt to changing internal and external conditions, and be life-supporting, creative leaders need to be effective listeners so that they can collect feedback and continually assess their systems. In class, students learn about different types and levels of listening, including 'attentive listening' which teaches students how to respectfully, thoughtfully, and effectively listen. Listening skills are practised in the context of 'Yes, and', Socratic Inquiry, and Collaborative Inquiry.

Creating Triads

Conventional leaders tend to create a series of one-on-one relationships in a network that looks like a radiating spider web. To maintain top-down control, conventional leaders seek to inhibit relationships between and among subordinates.

Being resilient, optimizing, and systems-based requires the development of strong dynamic interconnected networks, both within and external to the organization. Creative leaders develop strong resilient networks by creating and connecting triads (three-member relationships), each of which is based on shared values, and that generate value for all members through active participation in the triad and network (Logan et al. 2011). Students practice building values-based triadic relationships that have the potential to generate value for each of the members.

Examples of Emerging Creative Leadership

Six months after the course ended (when many students were still in the program) students in the pilot course were asked to share the leadership capacity they felt they personally developed from the course and if/how they have already used that capacity to benefit their communities. Their paraphrased responses are bulleted and grouped under selected Creative Leadership practices, principles, and characteristics that they reflect.

Becoming a Creative Leader

- I recognize that I can be a leader in any group by helping set the culture. I practised this with our (online) class, and the instructor noticed the change and thanked me for enhancing the class participation and learning experience.
- Before the course, I was not on good terms with what I thought was 'leadership'. The course helped me see that I am moving from stage to stage, to see in what stages I did not want to be, and how to move forward.
- I recognize I can lead by helping challenging people move up to the next cultural stage, rather than blaming or punishing them.
- I remember the feeling of a 'stage 5' experience and immediately resolved to make such a work environment part of my goal.

Being resilient, driving towards positive outcomes

- When working in a group at a lower cultural stage, I used to just get frustrated or annoyed, but now have a sense of what needs to be done, what can be done, and what I want to do to move the group ahead.
- I inadvertently initiated a highly confrontational political discussion regarding sustainable design in a challenging situation. I dreaded even continuing the conversation; however, using tools from the course, what would have been an ugly scene came out thoughtful and positive. I was very proud of how I handled that situation and can guarantee you that before the course I would not have engaged in this discussion.

Being values-based, systems-based

- I recognize how important it is to be true to yourself, even in a context that pushes the other way.
- I now realize that we feel naturally drawn to values we cherish such as humbleness, honesty, empathy, collaboration, etc. Without collaboration among individuals we would not be where we are. Individuals that help bring together objectives and collaboration amongst the group members are the natural leaders. That does not mean to delegate and give orders, but rather to listen, motivate and be an active member of the group, preach with the example.
- I now try to create triadic relationships as often as possible and am honing my abilities at building my tribe.

Being optimizing, adaptable

- I am a person of action, but I have learned to pause and reflect, get feedback from my community, then act with more respect and intention, in order to drive towards positive outcomes.
- I have shifted from giving top-down advice to showing empathy and using inquiry to help resolve problems.
- I relished the examples of those who had mentored and been mentored. That is something missing from the graphic design industry today.
- I learned that leadership models in nature are flexible in order to survive. This is a great knowledge I gained from the class. The ability to use different leadership models to engage different individuals and to recognize that there can be different models for different situations.

Conclusions

The goal of the Creative Leadership course was to empower emerging sustainable designers to become effective leaders and, in doing so, build the capacity of the communities with which they will work to achieve sustainability goals. In addition to building a base of knowledge and understanding of creative leadership, especially as compared to conventional leadership, the course included a unique focus on learning and practising methods for developing and expressing creative leadership values and characteristics. Due to positive feedback on the pilot, it is now a 15-week core course in the SDO program. Although becoming an effective community leader can be a lifelong journey, this course helped change students' understanding of leadership and themselves as leaders, and empower and equip them with methods to begin practising creative leadership within the communities with which they work, thus building capacity to reach sustainability goals.

References

Biomimicry for Creative Innovation. (2011). Nature's principles. BCI: Biomimicry for Creative Innovation. http://businessinspiredbynature.com/wp-content/uploads/2013/02/Natures-Principles.pdf. Accessed 24 May 2012.

Belsky, S. (2010). *Making ideas happen*. London: Penguin Books.

Bennis, W. (2009). *On becoming a leader*. New York: Basic Books.

Benyus, J. (1997) *Biomimicry: Innovation inspired by nature*. New York: William Morrow & Company.

Chakravarthy, B., & Henderson, J. (2007). From a hierarchy to a heterarchy of strategies: Adapting to a changing context. *Management Decision, 45*(3), 642–652.

Christensen, C. (1997). *The innovator's dilemma: When new technologies cause great firms to fail*. Boston: Harvard Business School Press.

DeLuca, D. (2012a). Ecological thinking for radical transformation. BCI: Biomimicry for Creative Innovation. http://businessinspiredbynature.com/wp-content/uploads/2013/02/Ecological-Thinking-for-Radical-Transformation.pdf. Accessed 11 June 2013.

DeLuca, D. (2012b). Leadership inspired by nature. BCI: Biomimicry for Creative Innovation. http://businessinspiredbynature.com/wp-content/uploads/2013/02/Leadership-Inspired-by-Nature.pdf. Accessed 11 June 2013.

DeLuca, D. (2012c). Rational radical. BCI: Biomimicry for Creative Innovation. http://businessinspiredbynature.com/blog/rational-radical. Accessed 24 May 2013.

Doyle, M. E., & Smith, M. K. (2001). Classical models of managerial leadership: Trait, behavioural, contingency and transformational theory. The Encyclopedia of Informal Education. http://www.infed.org/leadership/traditional_leadership.htm. Accessed 11 June 2013.

Logan, D., King, J., & Fischer-Wright, H. (2011). *Tribal leadership*. New York: HarperCollins.

Minneapolis College of Art and Design. (2012). Sustainable design. http://mcad.edu/sustainable. Accessed 20 Dec 2012.

Pascale, R., & Millemann, M., & Gioja, L. (2000). *Surfing the edge of chaos: The laws and nature and the new laws of business*. New York: Three Rivers Press.

Phillips, C. (2001). *Socrates café*. New York: W.W. Norton & Company.

Safian, R. (2012). This is generation flux: Meet the pioneers of the new (and chaotic) frontier of business. http://www.fastcompany.com/1802732/generation-flux-meet-pioneers-new-and-chaotic-frontier-business. Accessed 11 June 2013.

Slater, S., Mohr, J., & Sangupta, S. (2013). Radical product innovation capability: Literature review, synthesis and illustrative research propositions. *Innovation Management*. (in press).

Wheatley, M. (2011). From hero… to host. *Resurgence, 264,* 14–17.

Chapter 8
Connections: Academics, Architects and Community Pro-Bono Projects

Sarah McGann and Barbara Milech

Abstract This chapter proposes a model for offering pro-bono architectural design projects to communities in a fashion that facilitates capacity-building opportunities for three cohorts: academic researchers, early-career graduates and industry professionals. The model provides a structure for collaborative work between all three participants, one that enables developing community-based pro-bono design projects that address real community needs. The model is first described in general terms, then illustrated by a case study comprising a pro-bono residential design project, then discussed in terms of the implications of the case study.

The community-focused project that forms the case study was designed by an academic researcher, developed by a recent architecture graduate and supported by practice professionals who provided mentoring, technical back-up and professional indemnity support. In addition to these pro-bono contributions from all participants, the project was supported by a small $ 3,000 publishing grant that enabled the academic to hire the graduate as a research assistant. At the conclusion of the project, each participant in this transient pro-bono community found they gained considerably from the experience. The recently graduated practitioner gained strong client/community experience. The architectural practice gained a more experienced employee and made a contribution to community. The academic found a way to connect academic capacities with community while developing new understandings of the productive nexus between creative practice and academic research. Moreover, and not least, there was a gain for the client, who became a fourth participant in the capacity-development effect of the project.

Keywords Research · Architecture · Model · Pro-bono

S. McGann (✉)
Department of Architecture and Interior Architecture, Curtin University, Perth, Australia
e-mail: s.mcgann@curtin.edu.au

B. Milech
School of Communication, Culture and Creative Arts, Curtin University, Perth, Australia
e-mail: b.milech@curtin.edu.au

R. Tiwari et al. (eds.), *M² Models and Methodologies for Community Engagement,*
DOI 10.1007/978-981-4585-11-8_8, © Springer Science+Business Media Singapore 2014

Introduction

It is not uncommon for academics teaching in university architecture departments to be asked by individuals or community groups to undertake real-life design projects that address individual or community needs which, for lack of funding or expertise, remain unmet. Often these requests are motivated by a well-meaning (mis)understanding that there could be a quid-pro-quo arrangement, in which students receive experience/education in exchange for pro-bono design services. Such requests, often appealing (always an appeal), present challenges to academics, challenges that mostly lead them to shy away from potential opportunities. In this paper we propose a model for responding positively to such requests, one that takes into account the challenges and suggests the benefits that can accrue to all participants in such pro-bono design projects.

The challenges are several. The overarching one is that architecture is a regulated and registered profession, one that requires considerable professional indemnity when a design process is attached to creating a building or space or place. Generally, neither academics nor students are so indemnified. Moreover, academics are time-poor—in Australia's current university settings, most academics carry onerous teaching and administration responsibilities that leave scant time for research, the life-blood of any advancement they might hope for. Thus the time needed for mentoring students or graduates engaged in pro-bono projects is most often neither available nor rewarded. That is the second challenge. The third pertains to students: Masters and doctoral architecture students, along with recent graduates, may not be sufficiently trained for such pro-bono projects; more importantly, they may not yet have met the registration demands of post-university professional experience and institute examinations. And, last but not least, there are constraints on professional architectural firms whose contribution is often implied in such requests: their first priority is to ensure that their firms thrive, with further education of any junior staff involved in such a request coming as a second or third priority.

This chapter proposes a model for offering pro-bono architectural work, a model that addresses all these considerable challenges in order to imagine how academic researchers (staff, graduates and, perhaps, enrolled postgraduate students), working with professionals in the field, would be able to accept pro-bono requests in order to respond to community needs. The model was developed through a particular experience of accepting such a request (and taking it to a successful conclusion), and is informed by understandings derived from a particular mode of creative-production graduate research. It is a model for community-based pro-bono design projects undertaken through a collaborative triad of academic/graduate/professional working together to service important community needs.

The experience that informs this model is presented here as a case study. The process began with an appeal to one of the authors: a request to provide a pro-bono residential design for a recently paralysed young client. The academic accepted the project, despite the sorts of challenges outlined above. She provided the initial design concept, then obtained a $ 3,000 publishing grant that enabled her to hire a research assistant, a newly graduated student who had recently been hired by an architectural firm. The research assistant, supported by technical back-up and professional indemnity provided by his architectural firm, developed the academic's concept design. This deployment of academic and professional resources, together with pro-bono time contributions from all parties, resulted not only in a published research paper (as required by the grant), but also in an advanced design solution that met the client's needs so well that he proceeded to complete the building as an owner-builder.

The potential of this experience as a model for pro-bono community work that entails collaboration between academics and practising professionals is informed by a particular model of creative-production research that emerged in Australian universities across the 1990s. The impetus for universities to accept higher-degree research that entailed a creative-production component sprang from the limitations of postgraduate research programs at that time. Australian universities had thriving undergraduate programs in areas like architecture, design, creative writing, fine arts, film and television, and the like, but limited options for graduands in these areas, for the protocols which then pertained admitted such students into postgraduate programs, but only in ways that implicitly discounted creative practice in favour of 'traditional' research into the historical, cultural or theoretical underpinnings of creative practice. Put simply, research students in such areas were expected to produce a discursive thesis, a 'not more than 100,000 word' discussion of some art practice. All that changed across the 1990s in Australia (and elsewhere) as academics in relevant areas pushed for a form of Masters or doctoral research that issued in a thesis whose form included a non-discursive component (a novel, a building design, a documentary, a collection of paintings, and the like) set alongside a discursive component.

The effort for reform in university higher-degree-research practice was coherent in its energy and diverse in the voices that spoke up from across the many university disciplines engaged in creative practice. However, ideas for the shape of Masters or doctoral theses in disciplines imagining a research thesis that included creative 'texts' joined in some way with a discursive component (what came to be called an 'exegesis') were far from coherent. Some argued for the exegesis (as the adopted word implies) to be a document that explained or contextualised the creative production that formed part of the research thesis, a document that provided a statement of the intention, antecedents and underpinning theory of the creative component of the thesis. Others argued that this approach was bifurcated: that it diminished either the creative or discursive component of the thesis, and—worst—lost the chance of developing a research project where a problem was defined, addressed and resolved through two modes of inquiry, two languages (see Marshall 2010; Milech and Schilo 2004).

The case study that illustrates the model proposed here for pro-bono community work is indebted to the second of these two notions of creative-production research: the notion that research which accompanies production can/should be more than comment/history/justification; that production that accompanies research can be more than a product, can also answer a significant research question; *and* that working in two modes at one time—production and research—can provide powerful, innovative answers to a well posed question. That is, the case study presented here is inspired by a conviction that there is a similarity in the process of creating *both* components of a creative-production thesis—production and exegesis. Neither process is linear; both depend on a sense of structure. Both start from a core intuition/idea/answer-to-a-research-question that develops, after studio and library work (via sketches, concept maps, investigation of prior ideas relevant in the problem) into related frameworks for both production and exegesis (site sketch in architectural terms, a provisional chapter-plan in exegetical terms).

Those related plans are then tested and revised, again and again, through working between the frameworks and activities of production (the detailed plan) and research (reading/viewing/writing). In both instances—in work related to both production and exegesis—the process is circular/reciprocal/*creative*, a process in which frameworks are (continually) refined and details (continually) find a cogent place in both work and text. The special challenge and excitement for creative-production research conceived in this way is that the work takes place not only within one genre (production *or* text), nor only in two genres (production and text), but crucially *between* two genres (production *and* text). The result—in our experience—is a rich reciprocity that makes creative-production projects, at their best, potent contributions to the disciplines from which they arise.

That potency also applies to pro-bono architectural work that is aimed at real-life problems. Architecture is a diverse profession, serviced by a diverse education that generally comprises design activities (50 %), technology/science activities (25 %) and culture/theory activities (25 %). Across Australia, growing student numbers enrolled in architecture courses require an increasing number of research-active academics to teach and research across these diverse areas. Old patterns linger in architecture departments, in so far as there still remains a kind of 'divide' between academics deployed as 'traditional' (those who explain/contextualise practice from culture/theory/technology perspectives) and academics considered as 'non-traditional' (those whose presence attaches to professional and creative-practice successes). And old hierarchies linger… This is so not only because of the (waning) weight of tradition, but also because, given the long design and construction timeframe of any building project and the limited number of peer-review awards available to such projects, it is often difficult to meet university research targets.

The model proposed here considers the whole design process—from inception to completion—as potential material for creative-practice research that contributes to community needs. It seeks to bridge the gap between academic structures and community needs by using academic systems in conjunction with public/professional ones. In particular, it proposes an innovative way to service community needs by harnessing pro-bono projects to academic outputs (for example, a visual essay, a methods paper), using the resources of universities (the possibilities of the research-assistant system, small in- or out-house publishing grant programs, etc.) alongside the resources available to professional practices (indemnity structures, programs for mentoring new staff, etc.). Such pairing of practice sponsors with academic researchers to work on community projects not only has the excitement of working between 'genres' (between project production and text-based research publication, between institutions, between disciplines like architecture and health sciences, etc.), but also has the reward of meeting a community need. The case study that follows tells the story of one such pairing used in the development of this model.

A Case Study: Designing with Six Hats

Neil is a sporty, outdoorsy sort of bloke. He spends very little time at home, preferring to golf, surf, go to the footy and hang out with his mates. You can tell just by looking inside his house—a huge TV in a darkened lounge, filled with golf equipment, surfboards and motorbike gear. That was before. Neil now is a paraplegic, his house is inaccessible and his boys' toys are in the way. His home, previously an inwardly focused place to sleep and store, will become the centre of his world (when he can eventually get into it). Initially, accessibility is the key problem to be solved. To this end, an architectural advisor from a rehabilitation unit proposed formulaic adjustments to the bathroom to allow for wheelchair access. However, accessibility is only one facet of the problem. This environment, relatively unimportant before, now needs to enable and inspire Neil both physically and emotionally.

The house, a solid 1960s project home, was dark, pokey and bisected by a long narrow central corridor. The toilet, bathroom, laundry and kitchen were all unusable from a wheelchair. In fact, the whole 750 m^2 block from front lawn to back fence was totally inaccessible to Neil after his trauma. The long grass sat like an incarcerating moat surrounding the house. It became obvious to the client and carers that solving the problem of toilet and bathroom access was not nearly enough. Dissatisfied with the $ 5000 government-supported solution—to gut and re-equip the bathroom similar to a hospital-style disabled bathroom—the clients looked for a better solution.

And so Neil's sister (and carer) approached Sarah McGann, a friend and academic, and asked her to look at some design proposals. The project presented interesting design problems (envisioning living scenarios for a motorcycle accident victim) combined with logistical challenges (lack of funding, lack of time, lack of professional indemnity insurance). These challenges led, eventually, to developing the model proposed here for pro-bono architectural work that builds alliances between clients (with community needs), academics in the field of architecture, and professionals working in that field. At the outset, Sarah explained to her friend that to take on such pro-bono work would require a research output that would not only justify the academic's time but also (potentially) enable funding for research assistance. Neil and his sister expressed interest in this process.

Initially the client's thinking involved alterations and improvements to the existing house. Then, following a generous offer from the Construction Workers Union, a new house on the same block of land was considered. This led to a more holistic approach with even more possible living scenarios—a family home, a share-house, a rental property or a house and land subdivision. The notion of accessibility became implicit in the various scenarios but was no longer the single driving concept. Then, in the way that happenstance is part of creativity (in both production and discursive modes), Sarah happened to read Edward de Bono's *Six Thinking Hats*, and found a framework for her intuition that design thinking at its best is a progressive rather than linear process.

De Bono argues that the main problem with traditional thinking methods is confusion. 'We try to do too much at once. Emotions, information, logic, hope and creativity all crowd in on us. It is like juggling with too many balls' (2008, p. xi). His

Fig. 8.1 Neil's house: before (photo: David Robinson)

Six Hats method proposes that the thinker deals with one thing at a time so that emotion is separated from logic, creativity is separated from information, and so on. In this way each mode of thinking is clarified and has a greater chance to make a valuable contribution to the final result. In regard to design thinking, de Bono remarks: 'For design thinking, possibility is essential. Logical thinking likes to work with facts. Design thinking has to work with perception. The three most important things in design thinking are: perception, possibility, and practicality' (2000, p. 222).

The thinking about Neil's living space and its possibilities for his new life drew on this characterization of design thinking, and, particularly, on de Bono's Six Hat framework for solving the problems entailed in designing Neil's new space for hopeful living.

This framework, as it applies to architectural design thinking, can be briefly summarised as follows. White hat thinking—the factual, rational and neutral thinking mode—focuses on the given, for example, the existing site situation. Black hat thinking—the critical, logical and cautious mode—seeks to highlight the problems of the situation. Red hat thinking—the emotional, aesthetic and poetic mode—identifies (the client's) desires as they relate to the senses. Yellow hat thinking—positive, constructive and optimistic—looks for possibilities in the situation, is a mode of hope and dreams. Green hat thinking—the creative mode—proposes ideas for the situation. And, finally, blue hat thinking—the control or organizing mode—brings together all the information/ideas/desires articulated by wearing the other hats to form a design strategy, a solution. This process for Designing with Six Hats is illustrated below through a visual representation of how that process related to finding a solution for Neil's need for a home filled with ease and hope, a home fit for a future.

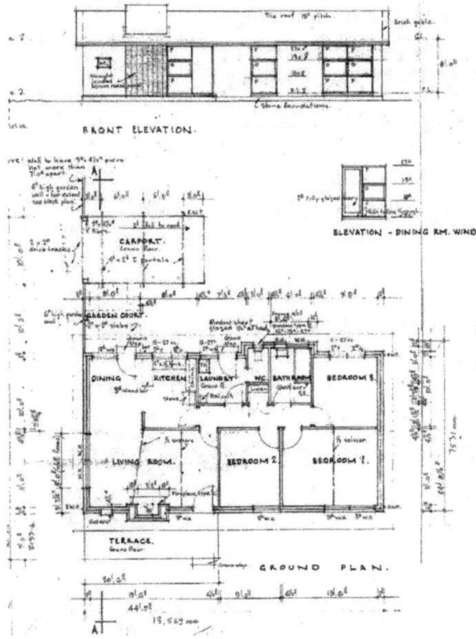

Fig. 8.2 White hat thinking: assessing the situation—factual, neutral

1. A small house in the middle of lawn and verge.
2. A solidly-build adaptable house.
3. A large subdivisible block.
4. Good neighbours. (McGann 2009)

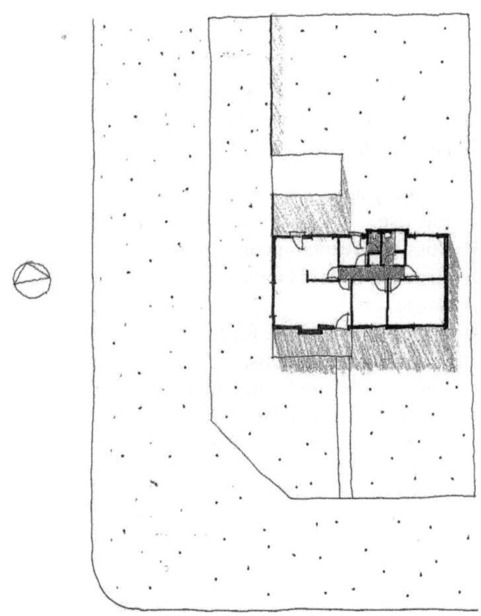

Fig. 8.3 Black hat thinking: problem-finding—critical, logical, cautious

1. Inaccessible house: bathroom, toilet, kitchen, bedroom, carport.
2. Inaccessible soft landscape: no access to house, no use for garden.
3. Poor orientation: no northern sun-light to house, no thermal delight, no sunny outside sitting area.
4. Poor outlook: no view to outside, no prospect, no visual connection to community. (McGann 2009)

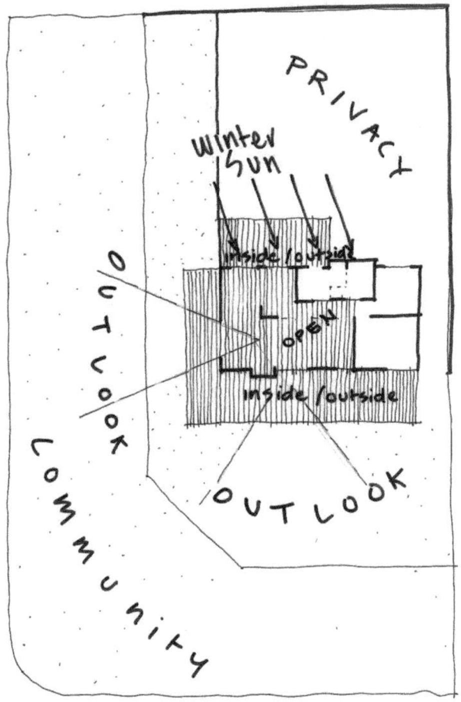

Fig. 8.4 Red hat thinking: perceptions—emotional, aesthetic, sensual, poetic

1. Hopeful: sunny, bright, warm.
2. Quiet: private and contemplative.
3. Lively: connected to community.
4. Welcoming: spacious and friendly. (McGann 2009)

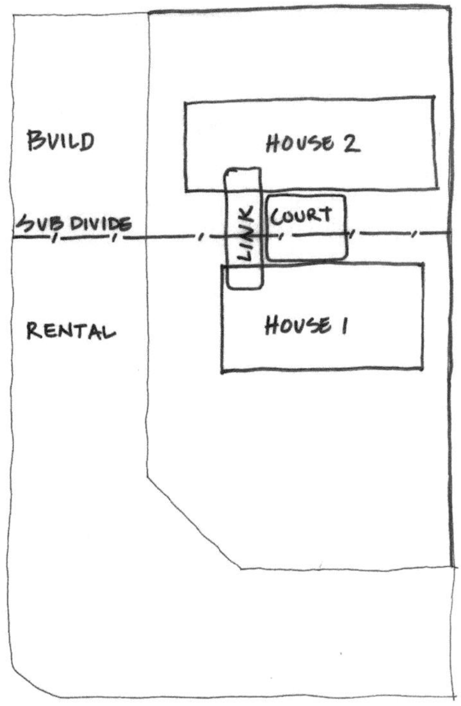

Fig. 8.5 Yellow hat thinking: possibilities—constructive, optimistic (hopes and dreams)

1. Possibly a large home with flexible use.
2. Possibly a rental income: a sustainable development of 2 houses.
3. Possibly a nest egg: for future subdivision decisions.
4. Possibly a share house: creating a community. (McGann 2009)

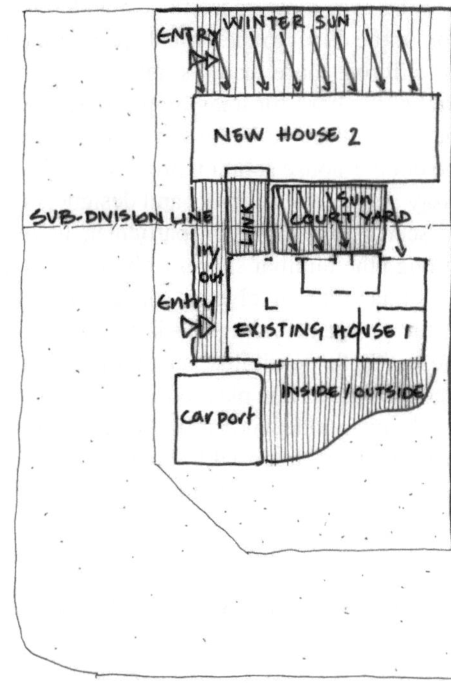

Fig. 8.6 Green hat thinking: propositions—creative, imaginative

1. Proposes a sustainable approach to development.
2. Proposes a courtyard house to maximise north light and garden connection, to provide a central focus.
3. Proposes verandas re-orientated to connect to the street.
4. Proposes decking and hard landscaping to fully utilise outdoors areas. (McGann 2009)

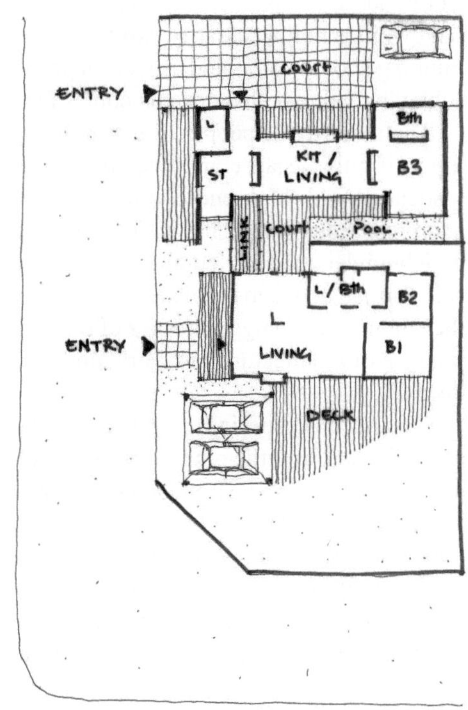

Fig. 8.7 Blue hat thinking: product—resolving, organising, strategizing

Final sketch design drawing. (McGann 2009)

What was most interesting in undertaking this project was the unexpected reaction of the client to the virtual spaces imagined in the ongoing architectural sketches. Offering more than a solution to an accessibility issue, the drawings provided Neil with a tool for imagining a space for his future life. From the presentation of the initial disabled-bathroom drawings by the rehabilitation unit to the sketches of the double-house solution provided at the end of the process, sketched in the visual essay above, client, carers and designers took a journey of discovery.

Scenario planning, in particular, helped the client and his carers to visualize, through the familiar spaces of home, how a new way of living could be carved out. This imagining involved a concerted yellow-hat effort and the temporary removal of the problem-finding black hat. The green- and yellow-hat modes combined to present both possibility and proposition. Possibilities, previously unconsidered, were discussed and planned for: a big party, a new family, a wheelchair basketball wind-up: an independent life. Even unbuilt, the sketches offered hope. So much so that the client proposed to be the project manager of the construction phases from his wheelchair: and why not?

Making Connections: A Model for Pro-Bono Architecture

Architecture is often described as a problem-solving profession: the bottom line always is that the design must practically address the problem and answer the brief. But good design ought to do much more than that: it must provide poetic and spatial qualities not necessarily explicit in the brief. In other words, design and design thinking should value-add as well as problem-solve. One of the prime difficulties in achieving value-added designs comes from the architectural profession itself.

The majority of people live in project-built houses that have little architectural input, if only because architecture is a profession generally only afforded by and associated with high-end commercial and residential projects, with large budgets and prestigious sites. That is, the benefit of architectural design thinking is neither widely understood nor equitably distributed. The model proposed here—a partnership formed between innovative architectural practitioners, research academics, and the community—suggests ways of remediating such limitations: it suggests that the practice of architecture can work alongside research to serve important community needs.

This need is especially significant for particular community groups, such as the aged, the sick and the rehabilitating, all of whom, in the context of shorter hospital and rehabilitation stays mandated by government and institutions, become more and more dependent on home and family as vital extensions to the healthcare system. In effect, home is called upon more and more to act as informal hospital/aged-home/hospice; and family are called upon more and more to provide primary care. This drift in current policy and practices for the care of the aged, disabled and rehabilitating in Australia creates a gap in regard to the built environment of those leaving hospitals, rehabilitation units and hospices to live at home; and mainstream project-built houses do not fill that gap effectively.

Thus, health-care trends provide challenges to the architecture community (both professional and academic)—some of which, as this case study suggests, could be met through a re-alignment within the architectural profession between practitioners and academics, working within opportunities provided by governmental and academic funding, to join together to invest in small-scale adaptation of existing homes *and* in holistically-minded development of housing solutions that meet special community needs. Such an alignment would work best when it provides a positive sense of place for those whose needs are not addressed by standard (project) architecture, when it provides a 'home' that is a place of ease and hope.

A negative visualization of place, like that generally associated with hospitals, rehabilitation units and nursing homes, undercuts hope, and hope is central to living. Place and identity are deeply connected: our lives are shaped by where we are born, where we live and where we die. That is, home is symbolic of who we are. This connection is never more apparent than in the misconception that a 'disabled' person needs a 'disabled' house. In *The Poetics of Space,* philosopher Gaston Bachelard describes the link between the notion of home and living, between home and a space for dreaming:

> The house we were born in is more than an embodiment of home; it is also an embodiment of dreams. Each one of its nooks and corners was a resting-place for daydreaming. And often the resting place particularized the daydream. Our habits of a particular daydream were acquired there. (1969, p. 15)

A final home with no space for dreaming (say, a standard hospital or aged-care facility or a minimally adjusted project house for a disabled person) cannot sustain hopeful living. And so, Bachelard suggests: 'Maybe it is a good thing for us to keep a few dreams of a house that we shall live in later, always later, so much later....' (1969, p. 61).

The model developed from the case study reported here imagines an alliance of professionals, working together within governmentally/institutionally dictated circumstances, to design houses for those in the community without access to 'high-fi' resources (see Till 2009), to build a home that we shall inhabit later, a place for dreams (and memories) carried forward from the past into a home that is good to live in.

This model suggests that all this could be possible, that innovative pro-bono architectural design services can be offered to communities through a process that deploys pertinent institutional resources in a fashion that facilitates capacity-building opportunities for all participants: academic researchers, early-career graduate and industry professionals (Fig. 8.8).

The model opens new possibilities for outputs in regard to both production (pro-bono design work) and text (academic research outputs). It highlights the importance of considering the architectural design process as a set of staged parts with each stage worthy of research output (a methods paper, a visual essay, a critical reflection), and suggests that low-budget architectural projects undertaken by aca-

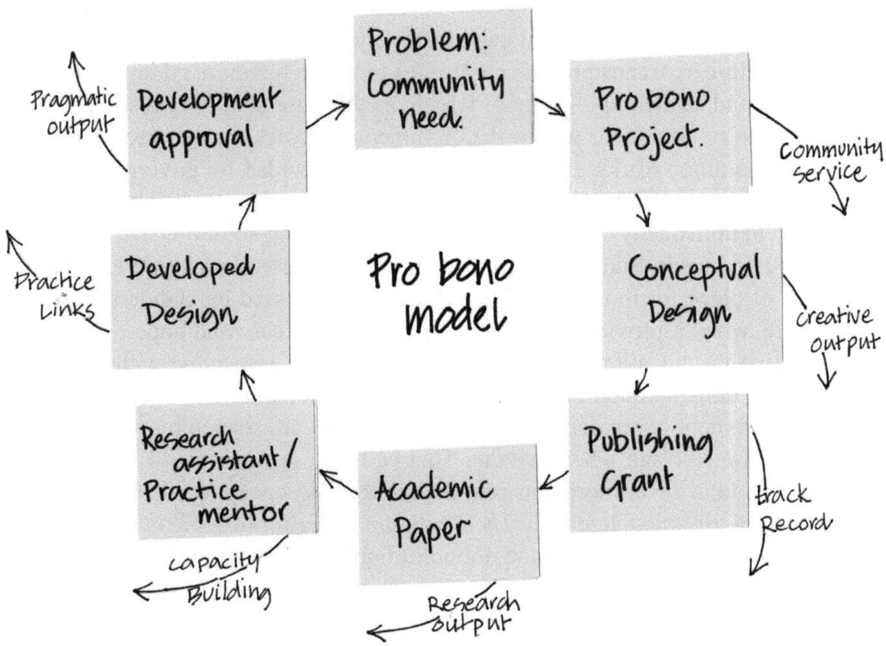

Fig. 8.8 Pro bono model diagram (McGann 2013)

demics can be poised to be more than 'just' creative-practice end-products. This requires a shift of emphasis for architectural academics and practitioners, both of whom, for different reasons, strive for architectural merit awards that generally are given only to high-quality and high-budget cutting-edge projects. It is unlikely that 'low-fi' pro-bono architectural designs like that provided to Neil could make rankings for 'quality' design awards. The great promise of this model, then, is that it suggests a way for both academics and professionals to be engaged in community-need-based architecture in a way that meets their separate needs as they come together in a potent alliance.

At the conclusion of the project described in the case study above, all participants found a considerable gain in capacity—the academic's capacity to publish, the professional practice's capacity to develop an employee's skills, the graduate's capacity to enhance his early-career progress, the client's joy in building a house that capaciously met his needs. That is, all the participants in this project met their separate personal/institutional needs. The client gained a design far beyond his expectations and developed ownership of the project through the consultation stages. The graduate/recent employee gained strong client/community experience, formed a bond with the client and grew to understand the everyday diverse needs of wheelchair users. The architectural practice gained a more experienced employee and made a valid community contribution. The academic gained currency in institutionally approved research, through the intermediate steps of grant, publication and creative-practice outputs. Each participant in this project gained something pertinent to their life circumstance and professional trajectory. Best of all, each learned something they didn't quite know before. That is the potential of the research model of creative-production research that is anchored in a conviction that research and production are not two separate processes, and that harnessing the two trajectories through a single statement of the problem to be resolved opens ways for new understandings, new creative projects, new solutions to community needs.

References

Bachelard, G. (1969). *The poetics of space* (trans: M. Jolas). Boston: Beacon Press.

De Bono, E. (2000). *New thinking for the new millennium*. Beverly Hills: New Millennium Press.

De Bono, E. (2008). *Six thinking hats*. Camberwell: Penguin.

Marshall, C. (2010). A research design for studio-based research in art. *Teaching Artist Journal, 8*(2), 77–87.

McGann, S. (2009). Switching hats: From accessibility to possibility. *Australasian Medical Journal, 1*(13), 162–166. doi:10.4066/AMJ.2009.134.

Milech, B., & Schilo, A. (2004). "Exit Jesus": Relating the exegesis and creative/production components of a research thesis. *TEXT,* (3), 1–13.

Till, J. (2009). *Architecture depends*. Cambridge: MIT Press.

Chapter 9
Fremantle on the Edge: A Community Collaboration

Anne Farren and Nancy Spanbroek

Abstract This chapter addresses community engagement presented through a large local, inner urban project held in the City of Fremantle in 2009. It was an ambitious project with many challenges. This is a timely reflection on the benefits and pitfalls of a project of this scale, which included a large community-based exhibition presented in sites throughout the city and a closing event that became an interactive performance engaging both exhibitor and viewer. The project was carried out within a very tight time frame of 5 months from conception through to completion, and required the coordination of a complex set of inter-related tasks associated with the securing of funding, liaison with key stakeholders and general management of the processes associated with working with community and the presentation of a range of events that, on reflection, represented a mini festival.

Keywords Student community engagement · Inner urban design project

Introduction

Fremantle has a unique local identity shaped by: its indigenous history as a meeting place for Aboriginal people during pre-European settlement; its geophysical location at the mouth of the Swan River; its history as the Western Gateway to Australia and the first point of arrival for many immigrants; its importance as a working harbour and fishing centre; its significant heritage buildings; its strong labour history; its arts, cultural and sporting activities, and its binding community spirit (City of Fremantle 2009–2010).

N. Spanbroek (✉)
Department of Architecture and Interior Architecture, Curtin University, Perth, Australia
e-mail: n.spanbroek@iinet.net.au

A. Farren
Department of Design, Curtin University, Perth, Australia
e-mail: a.farren@curtin.edu.au

R. Tiwari et al. (eds.), *M² Models and Methodologies for Community Engagement*,
DOI 10.1007/978-981-4585-11-8_9,

To meet the changing needs of Fremantle's City Centre, the City of Fremantle established an Action Plan in 2008 for the central city mall and streets, designed to enrich the existing social, cultural and business environment. Their vision was to create a precinct that was cohesive and coherent and that retained the unique sense of place of each of its public spaces. Our development of the Fremantle on the Edge project in 2009 was timely because the City was at a crossroads with a change of leadership, political divisions within the community, social problems and ongoing social changes within the community and local business. We were inspired by the potential that a large creative arts event developed over a period of years could provide, by profiling Fremantle in an international arena, and stimulating local business and community engagement.

This chapter reflects on the processes applied to the development and implementation of the project. It presents explorations into the design process, exhibition and the outcomes, and is contextualised through descriptions of the scope of the project and considerations presented to participating international artists and students. The community engagement achieved through the exhibition and performance stage and capacity building aspects of the project are outlined in the later part of the chapter. The conclusion provides a reflection on the project, its success and the overall outcomes.

The Project

Fremantle on the Edge was an interdisciplinary project involving 260 Curtin University students from a range of cultural and discipline backgrounds, which reflected the community of the City of Fremantle. The project involved City Councillors, international visiting artists, retailers and the general public, and engaged students in a work integrated learning experience. As professional curators and project managers we were instrumental in facilitating this experience.

The benefits of a project of this calibre were highlighted to the Council. They were perceived and presented as follows:

- This project would promote the City as accessible, responsive and forward thinking, actively engaging the local community in dialogue and interaction with the 're-visioning' process and re-examination of ideas about the city. This would be facilitated through interaction with artists and students involved in research activity to be carried out in the streets of the city, together with symposia, exhibition and displays over an 8 week period.
- The project outcomes would enhance the perception of the city as 'A beautiful, accessible and liveable city that recognises, protects and promotes its social, cultural and built heritage' (City of Fremantle 2011a, p. 33), in response to the City's Mission Statement.
- The unique character and strong sense of identity which exemplifies the City of Fremantle is embedded in its cultural history and diversity. This history would

inform the development of new ideas by participants in the project. As part of the presentation of their ideas, students were required to research and include the historical/cultural and contextual considerations that were significant to their selected sites. It was envisaged that this material would provide recognition of the cultural diversity and reinforcement for the strong sense of identity which characterises the Fremantle community. Additionally, it was expected that:

- The view of Fremantle would be enhanced through activities to draw attention to local creative industries such as architectural practices, fashion houses and artists, craftsmen, potters, sculptors, jewellery designers and art organisations based throughout the city;
- Students would engage in a re-examination of the city. Part of the process involved students researching the social, cultural and built heritage; exploring the nature of its beauty; and developing ideas which presented a potential new vision for sites in the City of Fremantle, with a focus on issues of accessibility and liveability;
- The visual material developed for the presentation of these ideas would provide a good database of ideas and materials that enabled the local community to see their city through fresh eyes.

The Mayor and councillors recognized Fremantle on the Edge as a good fit with identified community needs and supported the implementation of this project. A key value that was identified was the potential for this project to support the local business community, and much of our liaison with the City of Fremantle was through the Strategic Marketing Consultant who had been engaged to work on strategies that would enliven the West End of the city.

The Implementation Process

In retrospect, the implementation of this project can be seen as having occurred in two distinct phases: firstly, the 'Design in Context' phase was characterized by the 3 day workshop with the international artists, Maria Blaisse and Cocky Eek, and the 'Design Exploration' phase conducted at Curtin University and the Curtin Urban Sustainable Project (CUSP), which became our home base in Pakenham Street. This base enabled students to meet in groups, hold tutorial discussions, workshops and student presentations within the centre of Fremantle. This proved a convenient option, particularly for students working directly with the visiting international artists (Fig. 9.1).

Design in Context Phase

Students engaged in the initial 3 day intense workshop, with international artist and academic staff each assigned a group of students to mentor and guide through

Fig. 9.1 Fremantle on the Edge project: CUSP workshop presentations. (Photo Richelle Doney 2009)

their investigations. Students were assigned to groups randomly, regardless of their discipline or year of study. This later emerged as development of interdisciplinary skills within the student cohort. An initial forum was held to introduce the project, the format and how the outcomes would be exhibited. Visiting artists also made presentations on their 'European' perspective of the city, its site and its value.

The investigations carried out in the city by student groups during the initial 3 day workshop activity were presented to all the participants back at Pakenham Street base at Curtin University's Urban Sustainable Project (CUSP) as the final component of the introductory component of the program. Each group focused on different aspects of the city, ranging from the ephemeral to practical environmental factors, from light and shade to streetscapes. The closing seminar fueled a rich cross-disciplinary exchange of responses to the city environs and its cultural context. This was the entrée to the project, where ideas were stimulated, appetites whetted and energy levels high.

Design Exploration Stage

Following the 3 day workshop, students worked with their disciplinary leaders on the review and further explored the development of their ideas with a more specific discipline focus. As part of the design exploration stage, students were briefed to engage with the 'main course': the analysis of relationships between space and form, between interior and exterior space, between body and space and between spectators and performers, and the principles that organize them in our built environment. How we inhabit, engage with and occupy space are of critical importance in this new century, and an understanding of relationships between place, space, and its politics are indispensable for any sort of practice today. Occupation may be a state of mind where daily routines and activities are curtailed, molded and adapted to a particular environment. Equally, spaces may be composed or formed through processes and layers of inhabitation (City of Fremantle 2011b). Following the introductory lectures held during the first stage of the project, a series of talks was delivered by invited guests from visual arts disciplines from within the state. The talks were delivered to both the students and the public at a selected venue in the City. This process served to provide a base knowledge about the city, and perspectives of how the experts viewed the city from their respective disciplines. The talks also acted as stimuli for the exploration of ideas, and possible approaches to tackling the development of ideas within the project context.

As is typical of studio based learning, students regularly presented their work to other disciplinary groups, inviting positive critique and guidance on how to move forward. Each group's approach was different, thus inspiring students to reflect on their approach, experiment and discuss alternative approaches. Students were also curious about the progress and direction of others involved in the project, and we observed a level of friendly competition amongst groups driven to excel in their project outcome.

Setting the Scene

The design briefs given to participants were intended to act as a directional guide for students, allowing them to move into the final stage of their projects. Project participants, in this case primarily tertiary students, were asked to design and build models or full scale insertions that investigated the social, cultural and physical requirements of the space. Interior space can be that space between, the street space and the spaces beyond. They were required to consider points of reference, connection, framing and vistas, to understand the spatial qualities formed by insertions, controlled viewing angles, disguised forms and masked connections, and by qualities of filtered light, movement and shadow.

The participants needed to address real life situations, the pragmatic aspect of their selected site, natural light and security, and to gain an understanding of the users and uses through observation: their needs, their movements, their day and night activities. How is the street used; how does one respond to the shop front, office window, laneway and roadway? The shop windows became key exhibition sites offering a viewing spectacle both during the day and the night (Fig. 9.2). Baudrillard suggests:

Fig. 9.2 Love in Tokyo installation. (Photo: Lara Mackintosh 2009)

> Whether as packaging, window or partition, glass is the basis of a transparency without transition: we see, but cannot touch. The message is universal and abstract. A shop window is both magical and frustrating—the strategy of advertising an epitome. It is this immaterial materiality that ushers in a new theatricality to the city. (Baudrillard 2005, p. 42)

Students were invited to study the scale, the materiality of their environment and the existing light conditions, and to question how the buildings were being used, and how their site could be better utilized.

Participants studied the spatial organisation, the relationship of surrounding form and the space in between. Analysing the character of the site—the built form, textures and materials, the activity patterns and movement network in the area—what is the 'sense' of the place or its identity? Designs were to fully explore the relationship to the surrounding area, interrelationship of pedestrian movements, relationship between the spectators and performers, and the significance and shaping of the public domain. Students were asked to test and explore their intentions and assumptions so as to realize a convincing final design. Models and drawings were used in the process of developing an architectural proposition. Selected materials and fabric were employed to add to and enhance the poetics of space.

Presentation: The Event

> On a small strip of beach sitting on the edge between city and coast, guests watched white figures perform, silhouetted against the embers of fading summer light. Others 'played'

> with bamboo structures in the small limestone building edging on to the beach. As a cool evening breeze settled over the streets of Fremantle, small excited groups began to break away from the milieu that had formed at the front of Kidogo Art House[1] and move towards the centre of the city. Gatherings of intrigued onlookers could be seen clustered around patches of 'grassed' pavement or watching as ice hands melted away while others peered into retail windows. These were not casual window shoppers passing the evening away, they were guests at the opening of the exhibition. (Farren 2009, p. 71)

The presentation of works created by participating students became a performative event with works installed, inserted into a range of public places, private spaces and retail environments extending from Bathers Beach on the edge of the ocean and across the city precinct. A heritage wall in Cliff Street became the theatre for the projection of images narrating an 8 week investigation into the insertion of inflatable forms into the city and urban landscapes. High Street became host to a series of mini exhibitions presenting collaborative responses to the westerly winds, falling leaves, heritage textures and the smooth skins of modern buildings that are characteristic of the city. The empty windows into a local business showcased the veiled ghostly forms documented in the Roundhouse Tunnel that had once provided whalers access to the beach. The South Terrace promenade played host to poignant messages suspended from street trees and lamp posts. Components dangling from the branches of a tree invited passers-by to take a message home. Interaction with these works was allowed if not encouraged. The empty corner store that was once a lively fish market adjacent to the busy and bustling Fremantle markets was 'transformed into a maze of structures and images reflecting on the ebb and flow of the city nestled on the edge of the ocean' (Farren 2009, p. 72).

As the light faded on the opening evening events, steam and sounds echoing stories of the past escaped from a cardboard tram sitting on the fragments of track from a past era, and the performance was repeated one last time to cries for an encore. With map in hand, explorers engaged in an excursion through the displays, events and installations that mapped a journey into the city and its stories (Figs. 9.3, 9.4, 9.5 and 9.6).

'Architecture, rituals, festivals, businesses, domestic and market gardens, churches, industries, clothing, foods, history and peoples distinguish one community from another' (Kins and Peddie 1996, p. 10). The Fremantle on the Edge project culminated in a major event that presented a re-visioning of the local community through visual presentations of interpretations of the narratives drawn together by the students, and based on their experience of Fremantle—its people, buildings and spaces, all those elements that come together to define a culture.

Tangible community benefits resulted from this project for local business, the community and student participants. This community based project enabled students to engage in real situations, where the scale of the laneway, the street, the doorways and the shop fronts were tangible within the street as part of community life, engaged in activity, changing throughout the day and offering an accessible

[1] Kidogo Art House is located on the small beach, Bathers Beach, that defines the edge of the city of Fremantle and housed the bamboo works created by visiting international artist, Maria Blaisse for the duration of the *Fremantle on the Edge* exhibition.

Fig. 9.3 Fish shop Fremantle markets. (Photo Lara Mackintosh 2009)

Fig. 9.4 High street window. (Photo Ashley de Prazer 2009)

Fig. 9.5 Melting hands outside Myer Department Store. (Photo Lara Mackintosh 2009)

Fig. 9.6 Fish shop installation. (Photo Lara Mackintosh 2009)

and familiar framework to work in. Students were required to address real life situations—the pragmatics of their site, natural light, security, day and night street activities, together with identifying and observing the needs of the users of the area, and how they occupied their city.

The project contributed further to the reinforcement of the distinctive nature of the Fremantle community and the significance of the architecture, history, food, clothing and local stories to their culture.

Capacity Building

The project utilised a cross-sector approach, engaging businesses, council, educators, and the private sector. Community support was evidenced by the sponsorship response which was secured from 11 local Architectural and Design practices. In addition local businesses throughout the city offered their shops and offices as sites for the exhibition. An expression of interest in the project was received from the Fremantle Public Art Task Force, and representatives from the project were invited to give a presentation to a Task Force meeting on the outcomes and future potentials of the Fremantle on the Edge Project.

In general retailers were keen to see this project continued in the future because they saw it as an effective means by which to enliven the City's business district. This opportunity was of particular relevance to members of the High Street west end precinct. Other retailers not involved in the exhibition also expressed interest in being involved in future projects of this nature.

Eyewitness reports stated approximate numbers of public visitors to be in the order of 2,500 on the opening night activities. Tour groups were escorted throughout the city by academic leaders. There was general amazement at the quality and range of work amongst tour participants and visitors to sites over the duration of the exhibition.

A range of supportive responses were received from participating and supporting businesses, who generally felt the event was of great benefit to the community and should be repeated on an annual basis.

The Current Mayor of Fremantle, Brad Pettitt, stated:

> The Fremantle on the Edge project did make a great contribution towards the appreciation of arts in Fremantle. I think this can be put down to the innovate and creative way it brought art out of the usual gallery spaces and brought it to life it into the street and other familiar places that people would not normally expect to see art and engage with it. I also think its ephemeral nature really captured the public imagination and made people appreciate it and engage with it more as they know it may not be there when they next returned.

Past Mayor Peter Tagliaferrie stated:

> I have been involved in Local Government since 1983, first as an activist, a councillor and then as Mayor for the City of Fremantle for eight years until I retired in October 2009. In that time I have been involved in a period of immense change in Fremantle both social and in urban renewal, witnessing many significant events both on a national and international

> scale. Nothing interested or stimulated all my senses more than when I was made the presentation by Nancy Spanbroek and Anne Farren from Curtin University for Fremantle to host a pilot project for the inaugural Fremantle Biennale programme for 2012. What an opportunity, what a privilege. Biennales around the world are huge defining events for a city and a nation. (Tagliaferri 2009)

Conclusion

On reflection, more time for planning and development would have further enhanced the potential for a project of this scale. With limited resources, staff working on Fremantle on the Edge were stretched to their limits attempting to maintain their day to day duties at the University together with teaching support, administration, supervision, publicity, function coordination and security of the students whilst engaged on site. In future it is recommended that a similar event be held outside normal semester time, which would allow the students to fully immerse themselves in the project without the added distraction of having to attend to other areas of study, and the staff to devote their time to the complex supervision and management duties.

The involvement of international artists was key to the success of the project and of great benefit to the students. There is an opportunity here to engage local artists to participate in a project of this scale, to be directed by academic staff, to design and create their own street installation. The cross fertilization of ideas and design approach can still occur through workshops and presentations; however, the students would be better positioned to remain within their academic groups, ensuring academic direction by University staff. The project proved to be an outstanding and positive learning experience for students and staff, due to community engagement, and exposure to the internationally renowned artists Maria Blaisse and Cocky Eek.

Student projects in the disciplines of design tend to focus on hypothetical scenarios that isolate them from community engagement and testing of ideas in real world settings. Each project was unique and focused on a different aspect of the city in various sites. Students worked with their tutors in developing their ideas and, at the end of the day, presented to students and staff from other disciplines, providing students with experience in community consultation, public presentation, and teamwork. This stage provided broad feedback from staff, students and the international artists. Fremantle on the Edge became a vehicle for direct engagement with community in the process of capacity building.

Research indicates that students strengthen their learning of design and technical skills through collaborative work, where sharing of ideas in intense learning environments with multi-disciplinary groups informs their knowledge and broadens their skill base. Reflective practice is essential for problem-solving among students, and this occurs at its best when students work with students who have different skills, discipline-specific knowledge, and who challenge students' 'ways of doing' during the design process (Smith et al. 2005).

The project responses provide evidence that Fremantle on the Edge was extremely successful as a pilot project, demonstrating the potential for the establishment of a Fremantle Biennale of Creative Arts. For the successful implementation of a Biennale we suggested that the City of Fremantle appoint an Events Coordinator to manage the professional artistic engagement of local artists and the community. We also recommended support for the continuation of student involvement and staff input in order to retain the engagement of the new generation of designers in this process of re-visioning of community.

This project was created in October 2008, approved in February 2009 and exhibited in April 2009. The time frame was tight and the outcomes, although preliminary, revealed the impact a project of this size could have on a city such as Fremantle. Imagine what could be achieved with more time, a bigger team, and an international star list of artists together with our talented local artists, university staff and students.

While the project has not resulted in the allocation of funding to support the establishment of a major festival, the Council and community maintain the focus on the integration of festival activity in their community. The 2011–2014 Cultural Development Strategy for the City of Fremantle identifies four key strategic action areas: A Festival City, Infrastructure, Public Art and Heritage.

There is recognition of the economic value and significance of cultural policy: 'Cultural policy is increasing in a number of countries, extending beyond arts policy or heritage policy to embrace wider issues of cultural development and the role of culture in the national and international agenda' (Throsby 1999, p. 10). This reflects the establishment of generic community value in the support for projects that engage the community in reflection on and identification of its unique status. The Fremantle on the Edge project provided such an opportunity for a local community, as well as ongoing support for activities centred on community cultural development, to achieve the Council's vision 'for Fremantle to be recognised as a unique city of cultural and economic significance' (City of Fremantle 2010). Our project provides a case study for communities interested in building their community through the recognition and valuing of local culture.

References

Baudrillard, J. (2005). *The system of objects*. New York: Verso.

City of Fremantle. (2010). Strategic plan: 2010–15. http://www.fremantle.wa.gov.au/files/1818B384-4721-4F9A-91B8-9F8200AAE289/Strategic_Plan_2010_-_15.pdf. Accessed 6 Mar 2013.

City of Fremantle. (2011a). Annual report 2009–2010. http://www.fremantle.wa.gov.au/files/0f00aeb7-2cb8-4e9f-9db2-9e75010d297c/Annual_report_2009-10.pdf. Accessed 10 June 2013.

City of Fremantle. (2011b). Cultural development strategy: 2011–2014. http://www.fremantle.wa.gov.au/files/81c00fcc-faf5-478f-85c3-9f1800a86368/Cultural_development_strategy_2011-2014.doc. Accessed 12 Mar 2013.

Farren, A. (2009). Fremantle on the Edge: Personal Journal Entry (unpublished).

Kins, A., & Peddie, B. (1996). *Planning a complete community: A cultural planning guide for local government.* Perth: Community Arts Network.
Smith, D. J., Sanders, P. S., Demirbilek, N., & Scott, A. (2005, September). *Designing together: A collaborative experiment in design methodology within a multi-disciplinary environment.* In K. Holt-Damant & P. Sanders (Eds.), Third International Conference of the Association of Architectural Schools of Australasia (AASA), Brisbane.
Tagliaferri, P. (2011). Foreword. In N. Spanbroek & A. Farren (Eds.), *Pictures from the edge* (pp. 6–7). Perth: Paper & Pencil.
Throsby, D. (1999). Cultural capital. *Journal of Cultural Economics, 23,* 3–12.

Chapter 10
Transparency and Interdependence

Salvatore Di Mauro

Abstract It is argued that an inclusive approach to community-based projects is a sure way to guarantee its long-term success. While this paper supports this belief, it argues that the commissioned artist needs to see community as 'the crucible for positive development' (Goldbard, Arguments for cultural democracy and community cultural development Grantmakers in the Arts, p. 20, 2009). I believe that it is the responsibility of the artist to prepare the scene and identify a methodology that welcomes community engagement. The process in its entirety must be transparent and continually accessible. It is not enough to provide opportunities and avenues for bringing community into the collaborative team; the artist/project manager/art director must implement methodologies which build confidence and allow all participants an equal opportunity to contribute without bias. While confidence comes from awareness and familiarity, showcasing ongoing developments of the project beyond the collaborative team and out into the broader community will also work positively towards the success of the project, by strengthening community awareness and showcasing the skills of the participants. This chapter uses case studies to argue in support of this belief.

Keywords Confidence · Transparency · Showcase · Empowerment · Inclusivity · Context · Skills-audit · Collaboration · Narrative · Common ground · Sustainability · Ownership · Security · Mentor · Social capital

Introduction

It is commonly argued that an inclusive approach to community-based projects is a sure way to guarantee long-term success. While this chapter supports this belief, it argues that the commissioned artist needs to see community as 'the crucible for

S. Di Mauro (✉)
Queensland College of Art, Griffith University, Brisbane, Australia
e-mail: s.dimauro@griffith.edu.au

R. Tiwari et al. (eds.), *M² Models and Methodologies for Community Engagement*,
DOI 10.1007/978-981-4585-11-8_10, © Springer Science+Business Media Singapore 2014

positive development' (Goldbard 2006, p. 20). I believe that it is the responsibility of the artist to prepare the scene and identify a methodology which welcomes community engagement. The process in its entirety must be transparent and continually accessible; it is not enough to provide opportunities and avenues for bringing community into the collaborative team. The artist/project manager/art director must implement methodologies which build confidence and allow the participants an equal opportunity to contribute without bias. The methodology which I have devised has been directed by this belief, and I will discuss and illustrate the approach through a collection of case studies in this chapter.

Confidence comes from building on one's awareness and reciprocal understanding of a situation/need, and through developing an honest and sincere understanding of community and place. To this end, showcasing ongoing developments of the project, beyond the collaborative team and out into the broader community, works on many levels. Firstly, this strategy strengthens the awareness of the project and its focus; and secondly, it works positively towards featuring and showcasing those participants whose skills and local knowledge may not be known to the broader community. This exposure of project, and more importantly of the individual, has the long-term benefits of acknowledging their skills and suggesting opportunities for their ongoing contribution to community long after the completion of any project. Referring mainly to one case study—'Designing the Ideal Town' conducted in Yamaguchi Prefecture in Japan—with occasional references to other community-based projects which I have directed, I will argue in support of these recommended principles for community engagement.

Risks: The Artist

The risks that an artist who may be used to working independently in the secure haven of the studio has to come to terms with, when choosing to engage and collaborate with the local community, could include the following: loss of ownership of an original idea, compromise, loss of integrity in the work, people management, group/team work and direct dealings with the public. As well, there is a bottomless pit of possible liabilities and restrictions that mean one must surrender the security of self-expression for the sake of inter-subjective engagement; in all, one must part with the traditions of artistic object making.

As part of the process, the artist also needs to take on board the following responsibilities and embed them in his/her methodology: to listen and genuinely consider the needs and feelings of others; to empower others and genuinely direct their learning; to engage with politicians and funding bodies in an effort to gain support for the project; to hand over control—or at the very least work collaboratively—to generate positive informed decisions; to engage in negotiation with relevant personnel while managing time and adhering to deadlines; and to work alongside beginners as well as with the egos of other professional artists, administrators, and well-intentioned members of the broader community. Built into the process are also round table discussions, which rationalize and justify the process to the broader community while at the same time providing an opportunity for networking.

In this context, it can be said that the artist wears several hats, which may include: facilitator, art director/manager, educator, public relations officer, mediator, and above all, listener. To this end, it is understood that the artist may forfeit the quality of his or her work, and that intrinsic world principles need to be paired with strategic concerns and interests of supporters/sponsors from local industry and commerce if the project is to succeed in its entirety.

The Community Risks

Risks are many and broad for the artist. However, it must be understood that risks are a two-way street, and that the community is also taking risks, which could have detrimental effects on them in the long term. After all, the community is being directed by an outsider and the process of collaboration in its truest sense includes bringing conflicting sub-groups within the community together. This in itself can be a very unsettling experience for many, with the responsibility of spending local tax-payers' money (if local government supported) and working within budgetary constraints. Lateral thinking processes are everyday to the artist/designer, but may, for the non-art and non-design-initiated, generate feelings of insecurity and resentment and lack of confidence. A feeling of inadequacy, of not being knowledgeable enough to contribute, and with this a fear of exposing one's limitations to one's peers, can be a problem. Community-based art and design projects often, by their very nature (funding, community support etc.) require a serious commitment of time over an extended period until the project is completed, and beyond its completion once the artist departs the scene. This can be problematic for many people, as communities are quite often made up of transient populations as well as permanent, and this may result in ongoing changes in the collaborative team. While the more pragmatic prescribed method of running a business is what is required for its survival, business partners must also understand that the organic methodology of contemporary art practice must be considered as equally important if the project is to reach its desired outcomes.

The Artist's Possibilities

The advantages and possibilities are many, and to a large degree far outweigh the negative aspects of community arts projects, which are outlined above. In his or her capacity, the artist becomes the catalyst for the sharing of local knowledge. The artist moves from a process of being solely responsible for injecting an expressive content into the community to one of the community being proactive in the contribution of knowledge towards an unknown end.

The act of moving from studio to situation, as discussed earlier, has its long term benefits, 'diluting provocative assumptions about the relationship between art and the broader social and political world and about the kind of knowledge that aesthetic experience is capable of producing….' (Kester 2004, p. 9), and 'developing a relational aesthetic based around communication and exchange' (Bourriaud 2002, p. 113). The opportunity exists for the artist to showcase and encourage local knowledge and skills, and in the process provide sub groups with an opportunity to collaborate and contribute to their community. There is immediate and direct opportunity for the artist to get to know local culture through the locals, and in doing so, to develop a genuine understanding of place. By parting with the traditions of object making, artists adopt a performative, process-based approach. As Grant Kester (2004) finds, quoting artist Peter Dunn (2004), artists in reality become 'context providers rather than content providers' (Kester 2004, p. 1). Artists begin to see themselves as a resource for other people's non-artistic goals, such as local businesses, tourism etc. Quite often the creative thinker is asked to contribute his/her thoughts on some initiative being considered by the local business community which may be completely removed from any artistic pursuits but which require a lateral interpretation by a creative mind.

The Community's Possibilities

Community-based projects provide the opportunity for local citizens to come together in a non-political, non-biased way to share knowledge and understand each other better. Local knowledge is paramount for the outsider, and this can be used to strengthen awareness of place, and in turn, connection with the local citizens. Opportunities to highlight the abilities of local communities outside of particular areas of expertise or professional standing are few and far between, particularly in some rural communities. Creative art and design pursuits, especially if deeply reliant on an understanding of place, provide an ideal opportunity to identify and showcase knowledge and skills which exist within the local community but which quite often are never made public. Through such showcasing of events, employment and commissioning opportunities for artists and crafts workers are realized which may often lead to new financial avenues for the production of new art and cottage industries.

Frequently, dormant and obsolete skills are resurrected to support the production of the artwork, which results in positive mentoring roles between seniors and the youth of the district. In the process, local skills banks are updated and revitalized, and these in turn re-energize local industry, provide possible avenues for employment, and establish a culture of long term local sustainability.

The process, in short, is one of 'drawing on and the sharing of competencies as well as incompetencies—what collaborators know and don't know' (Deck 2004, as cited in Ang et al. 2011, p. 13).

There is great value in providing an open space for discussion, where participants can step out of their pre-existing roles and responsibilities in the community and freely express their concerns about place in an open non-threatening way.

The Qualities Which the Artist Brings to the Community

The nature of the creative artist revolves around questioning and reinterpreting. Schrödinger (1952, p. 109) argues that its not so much about 'seeing what no one else has yet seen', as much as it is about 'thinking what no one else has yet thought about that which everyone sees.'

Artists and designers are uniquely capable of generating the creative ideas that are needed to drive innovation. New ideas and the capacity to bring new ways of thinking to realization will be increasingly in demand in the future. Daniel Pink suggests that:

> We are moving from the Informative Age towards the Conceptual Age. From an economy built on the logical, linear, computer-like capabilities of the information age to an economy and a society built on the inventive, empathic, big-picture capabilities of what is rising in its place, the Conceptual Age. (Pink 2005, pp. 1–2)

In support of this argument, Kester makes reference to discussions which took place on Lake Zurich between politicians, journalists, sex workers, and activists from the city of Zurich:

> The group had been brought together by the Austrian arts collective WochenKlausur as part of an "intervention" in drug policy. Their task was simply to have a conversation about the difficult situation faced by drug addicts in Zurich who had turned to prostitution to support their habits. (Kester 2004, pp. 1–2)

This is a funding opportunity where art and health, especially mental health, combine for a positive result; both health and arts funding can be considered in situations such as this. Kester continues:

> Over several weeks dozens of these talks were organized involving sixty key figures: politicians, journalists and activists. Many of these would not have normally come together but in the ritualistic context of an art event, with their statements insulated from direct media scrutiny, they were able to communicate outside of the rhetorical demands of their official status and reach a consensus supporting a modest but concrete response to the problem. (Kester 2004, p. 2)

It is clear from this case study that one of the strengths that artists have, as an innate part of their skills package, is the capacity to laterally interpret and question a situation and to encourage others to do the same. The outcome is a creative act in whatever form it takes as it has moved on from bias and preconceived ideas into another world of possibility. In the Lake Zurich WochenKlausur talks, the 'art materials of marble, canvas or pigment were replaced by "sociopolitical relationships"' (Kester 2004, p. 3).

Artists and designers choosing to engage community must have a clear understanding of the focus of the community. They need to understand from the outset that they may be dealing with issues which they personally don't feel strongly about, but which nonetheless they must be prepared to listen to. It is important that the community feel a sense of ownership and connection to the project, and quickly realize that the role of the artist/designer is to facilitate the process, encourage discussion and direct an outcome. It is not the artist/designer's role to predict and preempt what the outcome will be. Operating in an atmosphere of true collaboration brings forth a comprehensive understanding of local needs, from which are realized a collection of priorities. To this end, the initial brief might be to create a solitary piece of public art, but through discussion, the perceived need might change from an art piece to an outcome with a stronger relationship between object, space and function. The artist must be careful not to overlook comments regarding perceived utilitarian needs from the participants and the broader community. Perhaps these needs can be combined and translated through functional utilitarian objects as well as artistic narratives of place translated through objects such as public seating, tables, shelters, bollards, drinking fountains etc. These items should have a strong artistic and narrative content but must also function on a multitude of levels.

Case Study: Ube Japan

Some time ago I was invited to work with a community in Yamaguchi Prefecture in Southern Japan. The local government had been made aware of how I was communicating my message of cultural conservation to the broader community through my creative artworks, and thought it would be interesting to observe how a creative artist would engage with the local community to solve a hypothetical urban planning problem.

At that time, I was travelling with a collection of 30 artworks through Japan.[1] My aim was to encourage visitors to the exhibition to look closely at their everyday traditions and how the originality of their place was fast being overtaken and homogenized by an internationalist, postmodern approach to lifestyle and place in the hands of careless urban developers (Di Mauro 1997).

The local government in Yamaguchi, as one of the sponsors of the exhibition, asked me to run a workshop with members of the local community. My responsibility was to pose questions that would result in the participants developing a better understanding of what makes 'an ideal town'. Their motivation for asking me to conduct such a workshop came from their understanding of my methodology for engaging and working with community. Perhaps they also saw the benefit of a non-local, non-Japanese artist/designer being involved, and of the possible attributes I could bring to their community.

[1] This exhibition titled *References* was a collection of 22 ceramic sculptures and eight collaborative artworks created during an artist in residency at Obora Pottery in Kyushu Japan.

The time allowed for the workshop was relatively short in relation to its aims and objectives; much was learnt in situ about working with such a broad mix of professionals and community members in a non-English speaking country.

Familiarity Works Both Ways

In retrospect, more could have been gained had I been better informed and better prepared for the challenges that lay ahead of me. Unlike projects undertaken since 1997, I went into the workshop with a basic understanding of Japan. This understanding grew from several years of visiting the country and several exhibitions of my artwork in various prefectures throughout the country. My personal vision of Japan was that of a cultural maverick with an unquenchable desire to know about this unique country and its people. This could be what drove me to accept the offer to conduct such an obscure workshop. Perhaps the officials had hidden agendas that were not communicated to me. Nonetheless it was a challenge, and one which I could only hope would not only satisfy the needs of local government but also provide the community with an understanding of the value of their comments and recommendations, when decisions about place and lifestyle were to be made.

Devising a Suitable Methodology

My teaching pedagogy revolves around drawing on the experience of my students and using this to energize further investigation. This methodology requires the teacher to be quite expressive in his/her delivery, and constantly aware of the emotional reactions of the students. In Japan this proved to be quite difficult and challenging because of the reserved nature of Japanese people. I was fortunate to be assisted by a professional interpreter, who was also both a highly respected member of the local community and a friend, who understood my way of thinking and my methods of working. This level of familiarity between artist and interpreter proved to be invaluable in helping to energize the community and the ultimate success of the project.

The task to develop a plan for the ideal town 'the town you want to live in' was presented to the participants via a collection of questions about their lifestyles and everyday needs. Many initially chose to listen and not speak. 'Japanese people tend to hesitate to stand up and talk in front of people, and the organizers had worried that there might not be any opinions or remarks coming from the participants' (Hirosawa, personal communication 1997).[2]

---

[2] Mrs Yoko Hirosawa was workshop coordinator and interpreter for the workshop, and her remarks quoted here and below are from a report she compiled for the local Yamaguchi council, and shared with me.

However, as the workshop progressed so too did their confidence and commitment, and before long they were all drawn into the task and engaging in free-flowing discussion.

The Significance of Place/Venue

While engagement happens as a result of the participants developing a feeling of confidence with each other, with the facilitators and with the topic in question, there are other issues that must be considered in the early planning stages of any workshop where a broad cross section of the local community will be participating. Kester (2004, p. 6) argues for 'creating an open space where individuals can break free from pre-existing roles and obligations, reacting and interacting in new and unforeseeable ways.' He highlights the need for common ground, for non-political spaces to be used for communities in an effort to break down barriers and to negate differences. In support of this argument, I reflect on the process which I used in the creation of the *River Reflections*[3] public artwork. The inaugural public gathering for the project took place in the Innisfail Catholic Church Hall under the shadow of the rather imposing architecture of the Catholic Church; perhaps this was not neutral ground for some of the residents! At such an early stage of the project, the hidden text of this location may have unintentionally ostracized many who felt intimidated by the prospect of participating in an event associated with what the venue represented. Some of the locals may have misinterpreted the project as being owned or at least directed by this denomination.

Kester (2004), in his pursuit of neutrality, also makes reference to community perceptions of select demographic groups within the community, and how they should be able to come together under the direction of the collaborative team to transcend the one-dimensional clichés promulgated by mainstream news and entertainment media. In this context, creativity truly becomes the umbrella of security, negating social, economic, ethnic, political and religious barriers.

The decision by the 'Ideal Town Project' coordinators and the commissioning body to conduct the workshop in a local, restored traditional Japanese family home instead of in one of the local government venues was a very positive move. The restored residence belonged to one of the founding fathers of the district, a highly respected medical doctor, his teacher wife and their children. After much thought, the family decided to restore the home and use it once again as a centre for cultural activities where local communities and others could come together to share cultures and lifestyles and learn about Japanese traditions (Fig. 10.1).

Since its restoration in the early 1990s, the home has been highly successful in achieving this goal. Music, poetry recitals, book readings, art and craft exhibitions and workshops with a local community focus have been staged on an ongoing, irregular basis. Once again, after many years of a silent existence, the Hirosawa family home has become a place where local culture and community engagement are

[3] *River Reflections* is a public artwork created by the author in 1998 for the Innisfail community in North Queensland Australia.

Fig. 10.1 Restored Hirosawa family home/gallery/cultural centre, Ube Japan (Di Mauro 1997)

celebrated. In this context, the venue provided the workshop with a much-desired neutral venue where inclusivity, as opposed to exclusivity, was encouraged and celebrated.

Familiarity and Confidence

Community-based workshops allow artists to become fully immersed in local culture. My initial connection with the locals in this instance was through the exhibition of my creations, which served as an icebreaker for those participants who were not familiar with my art. This, I believe, made it a little easier for the participants to want to engage with the process and with me, the artist. In a sense, the collection of artworks, and my visible interest and concerns for their place and traditions, became the necessary catalysts for connection. This sense of familiarity is important if one wishes to work with community, as it helps to break down barriers which locals may, subconsciously, feel between themselves and an outsider. For the artist, it means exposing who they are and how they think. For the creative practitioner, this is inevitable, and in my opinion, necessary as it provides the community with reference hooks which they can identify with and feel more comfortable and confident contributing to.

While, initially, I felt personally detached from this demographic, the project, without my intending it, preferenced immersion over detachment. Confidence in the Ube workshop was developed slowly at the outset, as the locals (from elementary school children to eighty-year-old men and women as well as some students from Korea and China) were invited to speak about their place and what it meant to them. Children, some as young as 10, were also asked to express their opinions,

Fig. 10.2 Open discussion following show and tell sessions in 'Ideal Town' project (Di Mauro 1997)

and were egged on by their parents, who felt a sense of pride knowing that their child's comments were being noted on the broadsheet at the front of the room along with everyone else. These process records provide the necessary visual reminders of the workshop focus; they were continually referenced, revisited and contextualized throughout the process (Fig. 10.2).

The art of engagement, according to art critics and theorists, is a kind of art which:

> … aligns with a strong trend in the international contemporary art world. A place where artists discard their posture of detachment in favor of deep immersion in the social worlds they find themselves in, and use this as the motivation for their artistic production. (Ang et al. 2011, p. 1)

This perceived trend is referred to in terms such as 'relational aesthetics' (Bourriaud 2002, as cited by Ang et al. 2011, p. 1) and 'from studio to situation' (Doherty 2007, as cited by Ang et al. 2011, p. 1).

Engagement

'Engagement is never a one-way street' (Ang et al. 2011, p. 1) but rather, a process which always elicits the participation and reaction of members of the community who come to the workshop from their own, often divergent, perspectives. Quite often, many come with an axe to grind. Discussions around the positive aspects of place inspired the participants to begin to open up about concerns and negative reservations they may have had about place.

As issues were brought forward, questions were asked in relation to the relevance and significance of perceived needs. Issues referenced included the loss of local traditions; for example, a rickshaw driver expressed his concerns about the fact that his industry was fast being seen as a tourist novelty and not as an ongoing aspect of Japanese culture. He also drew attention to how he and his family were deeply involved in supporting other traditional crafts. He argued for the role of traditional craft workers and how any 'new town' would need to seriously consider the implications of conserving what few traditions still existed in the community, and seek out opportunities for other traditions to be revitalized in contemporary society. Another key issue was the Westernization of the domestic architecture. Discussion here made reference to the traditional Eel houses of Hofu city in Southern Japan. The group believed that this highly functional style of housing could be used as a source of inspiration when designing domestic environments for restricted spaces. Other issues concerned the welfare of an aging population and the proximity of aged care facilities to social and medical networks. As Yoko Hirosawa puts it: 'the participants were very active and talkative. They expressed their opinions freely and worked positively to make their ideal town with great enjoyment' (Hirosawa, personal communication 1997).

Collectively unpacking the issues and needs ultimately revealed a suitable solution for each problem. Issues were contextualized, rationalized and prioritized, and progressively more visual data was added to the space.

With priorities and community needs realized, the group were asked to rationalize the data collected and provide the necessary facilities where sustainable lifestyles could be enjoyed and future needs accommodated. Following some debate and discussion, a list was made which provided a template for the next stage of the process.

With all the lifestyle needs discussed, and the infrastructure required to sustain a quality of life identified, the task was now to bring these together. Teams were formed of participants who could rationalize certain topics/areas from a professional and/or user need; for example, a healthcare group consisted of a doctor, a medical nurse, elderly persons and young mothers. Their responsibility was to discuss basic needs of medical centres and hospitals and their geographic location. Farmers, storekeepers and housewives discussed needs in terms of community farms, point of sale, access, parking and delivery. School teachers, children, carers and young parents discussed educational facilities—schools and childcare facilities. The professional was linked with the consumer and the indirect user of the space, and collectively they identified needs and priorities (Fig. 10.3).

Charged with their reference area, each group in turn presented their findings and sought approval and further recommendations from the participants. Again, here is a situation which provides an opportunity for people to speak with confidence from their collective experience. Slowly building this sense of confidence and empower-

Fig. 10.3 Presentation of concepts for the 'Ideal Town' project (Di Mauro 1997)

ment is crucial in any community-based project work. The groups presented their recommendations, which included the location of their facility in relation to other infrastructure in the layout of the ideal town.

Hands on Engagement

Perhaps the next stage of the process was the most dynamic, as it meant more direct and personal engagement. A collection of bits and pieces suitable for model making was provided, and it was from these that the model of the ideal town was created. Fun and excitement was the key, as the entire group worked diligently and with much gusto to create the three-dimensional model of their ideal town. The end product was photographed and featured in the local paper in the days that followed. The sense of achievement was evident on the faces of the participants as they posed for photographs with the model of 'the ideal town', which they helped to rationalize and in turn build (Fig. 10.4).

On Reflection

The methodology worked well to bring together a community of persons interested in understanding the needs of a better future for their cities. Through a process of confidence building and spontaneous and uncensored contribution, the locals felt an

Fig. 10.4 The model for the 'Ideal Town' takes shape (Di Mauro 1997)

affinity with the project in an atmosphere of sharing. As I understand it, The Ideal Town was not manifested in its entirety. However, I would like to believe that the methodology which was used to arrive at the desired conclusion has assisted the locals and local government alike to realize the benefits to be gained from engaging creative thinkers, who are able to take a very broad and lateral approach to solving problems, and above all to realize the implications and the benefits of informing and engaging the community in the decision-making process.

> Japanese city planning is often done by the government or public offices. People are accustomed to it, while the public sectors and companies tend to ignore their opinions. It was rather unusual for them to get together and discuss what kind of town they would like to live in. As a result, the participants felt the workshop was enjoyable and fresh. They thought they were able to have a valuable experience. (Hirosawa, personal communication 1997)

The Skill of the Creative Artist

The project provided an opportunity for artists to engage people with little or no experience in the arts, and tested the participants' confidence and faith in the artist. Though the timescale was challenging, to say the least, the process used remains paramount in the minds of the participants. While a political agenda may have been the motivation for the workshop (I will never know), my ignorance of this did not impact on my facilitation of the workshop; the social agenda far outweighed any possible political orientations. In the end the artist, with the community, has produced responses that demonstrate the power of embedding creative thought into non-arts practices. 'By keeping the artist in the center of the equation new ways of operating emerge, sometimes from unforeseen directions' (MacGregor 2011, p. ix). MacGregor's argument in support of the creative ability of the artist engaging with community is brought home in the following statement: 'Because of the way they are able to reorient our view of the world, artists are an underutilized resource that can be mobilized to help solve many problems faced by diverse sectors of society' (MacGreor 2011, as cited in Ang et al. 2011, p. 3). Case studies in support of this have been referenced earlier in this chapter.

> The workshop of Sam Di Mauro became an exciting experience to challenge new methods for us. It showed us a more active way of life with meeting people and opening up good discussion with them. From this experience we realized how important, and interesting as well, it is to get engaged in making our own town and community. (Hirosawa, personal communication 1997)

It is timely for artists and designers to realize this innate ability/power which they possess to laterally translate the world around them. But it is even more important that they see this as a skill (acquired or innate) to share with and pass on to others in an effort to arrive at a solution to a given problem, a solution which will positively impact on the sustainability of the community.

It is obvious from many creative dialogues that occurred in the process of community-based projects that the process from the very start until and beyond its end has the power to stimulate discussion and identify new directions for individuals and the community alike. And as such, it is less concerned with the 'formal condition of the object' (Kester 2004, p. 3) and more concerned with providing opportunities for all voices within the community to be heard and acknowledged.

All this leaves the creative thinker, artist and designer in a positive position to direct and inform the future and to extend their capabilities beyond the studio and out into the broader community. However, it's not that easy, and it's not every artist's desire to move along this path. The artist, in choosing to move in this direction, must be aware of the implications and needs that come as part of such a responsibility.

> I don't think artists are in any essential sense different from or better than other human beings. But our work—plunging our hearts and minds into the stuff of culture, attempting to see without filters or blinders, sharing the news with anyone who is ready to receive it—has special value in confusing times. (Goldbard 2009, p. 3)

The Qualities Embedded in Community and Which the Artist Brings into the Project

In an effort to extend the notions of ownership, it is important for the artist to seek out various skills that exist within the community or be prepared to train others to build up their skills in order to keep local content at its maximum. The Ulysses Link

Fig. 10.5 Kerry Scott, artist, working on the Ulysses Link piazza mosaic design (Di Mauro 2000)

Project (2000) at Mission Beach North Queensland provided such an opportunity for one of the participants, Kerry Scott, a newcomer to the district. Kerry's background was in graphic design. The opportunity to contribute to the project worked positively towards building her self confidence and hands- on skills. She created the main feature in the project, a 25-meter diameter mosaic that was installed in the piazza linking the costal walkways. This work is a testament to her determination to grow professionally and establish a career in the area of public art. She now lives in New Zealand and continues her public art practice through community engagement projects and private commissions. Other case studies will further support this argument (Figs. 10.5 and 10.6).

Quite often this retraining or reawakening of dormant skills continues to energize community and local industry long after the project has been completed. The River Reflections (1998) project situated in Innisfail, North Queensland, which I was commissioned to create in collaboration with local artists Rebecca Sweeney and Lily Heart, proved highly successful in this respect. When working on this project, we sought the assistance of a local potter and were fortunate to have found a couple of new arrivals to the district who had just set up 'Broken Nose Pottery' not far out of the town of Innisfail. The chief potter was commissioned to work on the project to produce a collection of feature tiles for the artwork. This position required her to tutor an indigenous group at the local Technical and Further Education College (TAFE) and in the process, produce some tiles for the artwork. This opportunity helped to broadcast the pottery to the broader local community, and has provided Broken Nose Pottery director Loraine Vagel with other

Fig. 10.6 Ulysses Link Piazza Mosaic Mission Beach by Kerry Scott 2000; porcelain, brass, stainless steel, glass, river stones, bushrocks, concrete. Dimensions of artwork approx. 25 m diameter (Di Mauro 2000)

commissions from local government on an irregular basis over the past 15 years. This, then fledgling industry, despite severe economic downturns in the craft industries, and two devastating cyclones, is still producing for the local and tourist market.

A Socially Inclusive Approach

Claudine Brown (2006, p. 11) argues in support of community based artists, stating that they value 'methods for building consensus, workable agreements and open, honest sharing. They strive for understanding amongst individuals with common goals, despite cultural and class differences, developing works that provoke discourse, stimulate participation and encourage action.' She believes that the creative mind of such artists often 'seek alternative ways of articulating and contextualizing issues during times of stress' (Brown 2006, p. 11). The sensitivity of the artist 'helps communities know themselves more fully, and can be surprisingly successful effecting profound change, such as reconciliation, where other methods have failed' (Brown 2006, p. 11).

We can glean from this that Brown has a comprehensive understanding of the broad scope of responsibilities that are part of the community artist's portfolio. Her descriptive analysis relates well to the tasks that are considered standard practice for the artists wishing to work with community. But she takes the description to much greater depth, making reference to not only the innate skills that come as part of the package that is the creative artist, but the sensitive and emotional responsibilities that the artist must bring to those situations where diversity is present and inclusivity is paramount. However, identified and articulated in this way, these responsibilities may ultimately generate reservations in the mind of some professional artists about committing to such collaborative processes, and as a result, fear of the unknown may deter them from accepting community-based public art commissions.

The Cultures of Place

Writing on culture, Bennett (2005, p. 63) states: 'The boundary lines between what is to count as national culture and what as regional culture have become increasingly contested'. However, he argues that there is also an

> extended range of users relating to forms of difference that operates within nations and across the relations between them: gay culture, lesbian culture, trans gender, ethnic cultures, diasporic cultures, and transnational cultures. The strong association that has been established between the concept of culture and the notion of lifestyle has generated another range of extensions from sub-cultures and counter cultures to club cultures, street cultures and drug cultures. (Bennett 2005, p. 63)

An opportunity may also exist for new arrivals to the district to be approached and encouraged to participate, thus breaking their apparent self-imposed silence. To this end, artists and designers have a social responsibility to acknowledge and provide opportunities for the engagement of all groups in the community.

As the artist prepares to enter the community, it is desirable to have a basic understanding of vernacular culture, including the sub cultures. This in turn will provide some understanding of how these communities are perceived and supported. These avenues of participation are most successful if and when the collaborative process involves cross-links between the collections of sub cultures (Di Mauro 2012, p. 477).

The Collaborative Process

Sociologist David Brain (2004) refers to community as a 'collective action by a group of people with common interests'. As a community they have an opportunity to voice their opinions and be heard—a confidence building part of the process. He argues, 'it's an act, which is performed together, and not just a container.' He believes that what makes a community 'work', in every respect is 'its culture and its governance.' 'The key component parts of the dynamic that must develop between the artist/designer and the community include shared understandings and expectations that people have of them, each other, and their collective endeavors.' These, he argues, are the things that make it possible for 'people to work together'. (Brain 2004, as cited by Borrup 2009, p. 4). 'Engagement is never a one-way street; it always entails the participation and reaction of multiple partners who come to the party from their own, often-divergent perspectives' (Ang et al. 2011, p. 1).

Collaboration between artist and community is a valued component of the public art process and one which results in a stronger connection to and ownership of the art and design work by the broader community. Collaboration opens up many opportunities through a process of sharing.

Process Not Outcome Driven

While the outcome may be different, the key element within each project is the creative facilitation of dialogue and exchange between all participants, their supporters and the broader community. Ultimately, the artist must be conscious of providing, throughout the creative process, opportunities for all to develop their self confidence. This will encourage spontaneous and sincere contributions to be made out loud, and for all to respond to. In this way, the process conversations are part of 'an active generative process that can help us speak and imagine beyond the limits of fixed identities, official discourse, and the perceived inevitability of partisan political conflict' (Kester 2004, p. 8).

These process conversations are integral to the finished artwork, thus making the two inseparable in the mind of the community. In a sense the finished project in whatever form it takes has a dual role: to act as a physical reminder of the way the community came together, and to celebrate and acknowledge place in a very public space for locals and visitors to enjoy.

Through the process of collaboration and through contribution to the creative process, a greater sense of ownership is achieved which will ultimately result in stronger networks being established. These collectively could identify and create other possibilities for collaboration.

The process is what ultimately brings the community together. The methodology that the artist chooses to apply is based on a clear understanding that the end product will evolve through a process of ongoing engagement and dialogue with the key players, the collaborative team, their support network and the broader community. Perhaps this is the key point of difference from the traditional creative arts approach chosen by the majority of professional artists.

Building Confidence

The collaborative hands-on approach tends to be an energizing force for all involved. This happens in a diversity of ways, frequently or occasionally, depending on the nature of the project and the methodology employed by the commissioned artist. While large groups of community being directly involved can at times result in very difficult situations to manage, it is the responsibility of the artist to seek out or at least be advised about the skills and knowledge base existing in the community. From this he/she is able to identify possible pathways where participants can contribute with confidence, and those without the knowledge base can develop their confidence to the point where they feel secure enough to communicate their concerns and recommendations. In the words of Goldbard:

> in community cultural development practice, participants' experience of their own creative imaginations and expressions is understood to be intrinsically empowering. As a practitioner describes: everybody brings something to the table and we need to help people figure out what that is, so they can have ownership. (Goldbard 2006, p. 54)

All those interested in participating in the process should be given the opportunity to experience work created with skill, ambition and intention. Each should be encouraged to express themselves and their concerns through their contribution. Each in turn should be judged by criteria appropriate to the intention of the work and not against some preconceived formula which constitutes so called 'good art.' As many community cultural development projects are built around learning experiences, the commissioned artist also works in a mentoring capacity to help the local artists develop and refine their ideas if and when necessary. It goes without saying then that the quality of engagement with the commissioned artist, local artists and the general public can positively alter the quality of the end product.

Transparency

The transparency of process, the understanding of the various stages of the development of the project, can be the key to successful outcomes including ownership. From my experience, transparency begins at the start of the project and continues to be reinforced throughout the entire process in various ways.

It is imperative that linking back to the local citizens remains a major and ongoing part of the process. Initial gatherings revolve around issues of connection to place, while subsequent works-in-progress exhibitions/displays build on the community's understanding of process and method. These exhibitions provide an element of transparency and are not only in an ideal opportunity to inform the locals of the methodology and its logic, but also to showcase those citizens who by choice are directly involved. They can be informal events staged in local community halls or, as was the case in the *Ulysses Link* project, illustrations of suggested public artworks were suspended from twine strung between trees at the site. Needless to say these events must employ every possible means of communication to guarantee the success of the occasion, as it is at these that the local community feel a sense of connection and empowerment. 'This is an opportunity for the community to voice their opinions and be heard, a confidence building part of the process. Not all situations provide an opportunity for showcasing to the community' (Di Mauro 2012, pp. 474–475). The *Q150 Mosaic* project was a case in point. Here the community spread over an entire state of Queensland and as such it was impossible for all to travel to one site to view the works in progress exhibition. In this instance the commissioning body, The Premier's Department of the Queensland Government, was able to attract the interest of ABC Television and in particular, Stateline, to the project.[4] A temporary exhibition was set up in a gallery space. The television team filmed the installation and interviewed the Art Director of the project. The segment was screened on statewide television in prime time and in this way many participating communities were able to see their proposals alongside proposals from other regional communities. This proved to be an invaluable part of the process and worked to bring the 73 regional communities together, albeit long distance.

The emphasis in such showcasing events is placed on the creators and the collaborative team; on how confident they are about publicly acknowledging their commitment to the project and showing and discussing their work with the broader community. Opportunities for showcasing the participants and the progress of the project throughout the process should be identified and used on an ongoing basis to inform the locals.

These works-in-progress exhibitions not only draw attention to the project but also become a celebration of the collective identity of the participants. They also remove any preconceived ideas about what may be misunderstood by the general public as secret business. They provide an opportunity for ongoing dialogue between the artist and the general public, and as mentioned earlier they showcase local talent and skills. All are very positive attributes when working with community and ones that, generally speaking, generate and encourage a sense of ownership and longevity for the project beyond its completion, and long after the artist/mentor/art director has left the place.

Community projects that expose process and product to the broader community at regular intervals throughout the project work positively towards helping to consolidate relationships between sub groups. In doing so they challenge preconceived stereotypical perceptions of other. The United Nations Universal Declaration of Human Rights published in 1948 states that 'everyone has the right to participate in the

[4] More information about the Stateline segment can be found at the following website: Q150 Mosaic Project Queensland's 150 Celebrations Stateline ABC Broadcast: 03/04/2009. Reporter: Jenny Woodward. http://www.abc.net.au/news/2009-10-30/cq-mosaic-images-go-online/1123004.

The Q150 Mosaic is a public artwork initiative that connects communities across Queensland Australia. Queensland councils were invited to provide an artistic representation of their communities. One hundred and fifty tiles were distributed throughout the State—two tiles for each council. In consultation with the community, local councils coordinated representatives from children through to noted artists, to embellish the "tiles"—one expressing the region's past and the other their aspirations for the future. The concept design and the individual mosaic tile shape were created by Queensland artist Elise Parups.

Once linked together this 'tessellated' geometric mosaic represents Queensland's history and aspirations for the future. The local artwork depicts stories, people, culture, history, aspirations and plans. The mosaic was built on site and installed at the Royal National Association (RNA) Showgrounds. The combined artwork was curated by Sam Di Mauro, Margaret Rackemann, Michelle Walker and Tamarra Rosman, and features 136 mosaic tiles from participating Queensland councils. The completed art piece has been gifted to the RNA by the Queensland Government.

cultural life of the community' (as cited in Goldbard 2009, p. 50). In support of this, Rene Maheu, Director-General of UNESCO in 1970, argued that:

> if man has the right to share in the cultural heritage and cultural activities of the community—or rather of the different communities to which men belong, it follows that the authorities responsible for these communities have a duty, so far as their resources permit, to provide him with the means for such participation. (Maheu 1970, as cited in Goldbard 2009, p. 50)

Social Capital

How does this impact on the role of the artist/designer? As stated earlier, the artist arrives in a community (physically or virtually) in most cases unaware of the skills and support that exist there. By getting to know the community the artist is in a sense empowered and able to realize potential for both short and long term engagement. For example, if the artist is informed about what skills and manufacturing facilities are available and accessible within the community, then he/she may be able to develop more meaningful pathways for collaboration. These industry pathways can have a two-fold flow-on effect and boost the local economy in the short term. In the long term they can provide educational programs and work towards improving/broadening and building on the skills base of the town.

With many of these skills, small-scale cottage industries can be set up which make locally identified product for local retail outlets. As well as economic gains there are also gains in social capital. These include the revitalization of dormant skills and in turn local industry. Along with this, individuals develop a sense of contribution and self worth, resulting in a more wholesome and contented community. In support of the argument to 'make local buy local', this is a philosophy which can ultimately energize communities and reduce the carbon footprint of the proposed art/design work. To this end, the culture of 'getting it made cheaply off-shore' as the best and only option, must be turned around.

Communities understand the needs of living in these environments and could work well and ultimately contribute positively to their rural wellbeing. The question is how to generate a sense of belonging within the migrant and the local communities, and how to encourage connection and contribution by new arrivals.

Global patterns are in a state of dramatic change; sea levels are changing; communities are being displaced; and the destruction of rural habitats is forcing mass migration to larger, already overpopulated cities. The collaborative team and in particular the commissioned artist must be aware of this dilemma and encourage participation in the project by migrant minority groups. The public artist/designer is in an ideal position to welcome and showcase the traditions and craft skills of newly arrived migrants and/or refugees. Perhaps through creative practices many barriers can be dissolved and removed. The art of place is for all the community and as such must be inclusive in its representation (Di Mauro 2012, p. 479).

Ongoing Contribution to Community

Naturally, projects, communities and their perceived needs differ in many ways. Some are about catalysing a consensus around a particular topic. In such instances there may not be any apparent physical outcome apart from an ongoing dialogue between user groups.

Through collaborating with community and ultimate ownership of the *River Reflections* artwork by the community, local government was able to gauge the success and long-term benefits of such collaborative art/design based projects. This awareness and ultimate success of the project was the catalyst for a further large-scale town beautification program, which took place in the years that followed. In this instance, several designers and creative artists were commissioned to contribute to the program, and again, many successful ongoing connections were established within the local community between artists, craft practitioners and trades personnel. Adding to this, several fledgling industries were boosted, while skills that had been dormant for some time were revitalized. It could be said that as a result of the success of *River Reflections,* local policy was modified to provide more support for the local visual art and design practitioner and his/her potential to boost the economy through local industry (Di Mauro 2012, p. 480).

Conclusion

While it seems that the responsibilities and risks are great, great satisfaction and rewards are derived from employing an inclusive methodology that shifts the artists' traditional process of working in the safe isolation of their studio out into the broader community, to share and showcase the scope of their abilities. This cannot only be rewarding emotionally and financially to the artist, but also provide an opportunity for sharing their skills/knowledge with others.

All this said, the creative arts practitioners must have the desire to be part of the community; also, as argued by Goldbard (2006, p. 20), artists need 'to see themselves as the crucible for positive development.'

With this understanding, the artist is well underway to achieving a successful collaboration. Yes, the risks are great and require some serious thought and consideration, but the possibilities and long-term gains are many and far-reaching for all parties concerned.

I will end with a quote from an interview between Goldbard and a practitioner, which I feel epitomises the desires of most community public artists:

> Everybody brings something to the table and we need to help people figure out what that is, so they can have ownership… [T]he arts allow us to imagine how the world could be different. … Quality involves the project leader's willingness to take risks and create partnerships that don't result in easy dialogues–real border-crossing. And they give a lot of credibility to the ideas participants are bringing into the project and provide a lot of tools to participants, so at the end they can make a space for themselves. (Goldbard 2006, p. 54)

References

Ang, I., Lally, E., Anderson, K., & Mar, P. (2011). *The art of engagement: Culture, collaboration, innovation*. Crawley: UWA Press.

Bennett, T. (2005). Culture. In T. Bennett, L. Grossberg, & M. Morris (Eds.), *New keywords: A revised vocabulary of culture and society* (pp. 62–63). Malden: Blackwell.

Borrup, T. (2009). *The creative coommunity builder's handbook how to transform communities using local assets, art, and culture*. Saint Paul: Fieldstone Alliance.

Bourriaud, N. (2002). *Relational aesthetics*. Dijon: Les presses du réel.

Brain, D. (2004). *Placemaking and community building*. Presentation at the University of Miami School of Architecture, Coral Gables, Florida, March 2004.

Brown, C. (2006). Foreword. In A. Goldbard (Ed.), *New creative community: The art of cultural development* (pp. 11–12). Oakland: New Village Press.

Deck (2004). *The Art of Engagement*. Crawley: UWA Publishing, 2011.

Di Mauro, S. (1997, June). *References*. Exhibition. Kyushu, Japan.

Di Mauro (2000). People Connection Place, Innisfail.

Di Mauro, S. (2012, October). *Community ownership*. Paper presented at Unidcom/Iade's 6th International Conference's Senses & Sensibility, Lisbon. (pp. 470–489). Lisbon: Ideiagrafix Design.

Doherty, C. (2007) Curating wrong places or where have all the penguins gone? In O'Neill, P. (Ed.), *Curating subjects*. London: Open Editions.

Dunn. (2004). Conversation Pieces: Community and Conversation in Modern Art. Berkley: University of California Press, 2004.

Goldbard, A. (2006). *New creative community: The art of cultural development*. Oakland: New Village Press.

Goldbard, A. (2009). *Arguments for cultural democracy and community cultural development. Grantmakers in the Arts*. http://www.giarts.org/article/arguments-cultural-democracy-and-community-cultural-development. Accessed 21 August 2013.

Kester, G. H. (2004). *Conversation pieces: Community and conversation in modern art*. Berkeley: University of California Press.

MacGregor, E. A. (2011). Foreword. In E. Lally, I. Ang, & K. Anderson (Eds.), *The art of engagement: Culture, collaboration, innovation* (pp. v–xi). Crawley: UWA Press.

Pink, D. (2005). *A whole new mind: Moving from the informative age to the conceptual age*. Crows Nest: Allen & Unwin.

Schrödinger, E. (1952). Are there quantum jumps? Part I. *The British Journal for Philosophy of Science, 3*(10), 109–123. doi:10.1093/bjps/III.10.109.

Part II
Unfolding Challenges and Removing Barriers in the Community Engagement Process: Opportunities for Transdisciplinary and Translocational Applications

The second part of M^2 explores the contextual evolution of community-based projects. It is critical that we, as community practitioners, reflect on our potential actions, our current practices, and our past ways of conceptualizing and enacting projects, as well as our individual and collective roles. The authors offer their processes and their learnings as a means to model this reflective mode of operating. Before we commence a project, we can gather together lessons learnt in order to ask: what may occur? What may be the consequences and impact of our actions? The cautionary note is that although projects or communities may be similar, each is unique and needs to be planned with due diligence, and evaluated as it proceeds and when it is completed.

> We shall not cease from exploration
> And the end of all our exploring
> Will be to arrive where we started
> And know the place for the first time.
> T.S. Eliot, *The Four Quartets*

Chapter 11
Capturing the Diversity and Commonalities of Community Engagement

Reena Tiwari, Marina Lommerse and Dianne Smith

Abstract Models and methods of engaging communities are greatly influenced by scale and the type of project, by the nature of the communities that are shaped by their locale, and by the disciplinary perspectives of the researchers. Are there any common threads that can be untangled from this web of diversity within which researchers, communities and projects operate?

This chapter provides a visual chronicle of all the community engagement narratives that have been presented in Part I. Researchers (authors of Part 1 chapters) workshopped together to compare and contrast the uniqueness of the approaches, models and methods used during the course of their projects. The aim of this chapter is to explore and present the diversities and commonalities embedded in these approaches.

Keywords Ethnography · Hybrid practice · Rapid participatory appraisal · Creative leadership

R. Tiwari (✉)
Departments of Urban and Regional Planning, and of Architecture and Interior Architecture, Curtin University, Perth, Australia
e-mail: r.tiwari@curtin.edu.au

M. Lommerse
Marina Lommerse Consultants, South Fremantle, Australia

School of the Built Environment, Curtin University, Perth, Australia
e-mail: marina@marinalommerse.com

D. Smith
Department of Architecture and Interior Architecture, Curtin University, Perth, Australia
e-mail: dianne.smith@curtin.edu.au

R. Tiwari et al. (eds.), *M² Models and Methodologies for Community Engagement*,
DOI 10.1007/978-981-4585-11-8_11, © Springer Science+Business Media Singapore 2014

Creative Communities: Inclusive Decision-making

Introduction

This chapter proposes some common threads from the diverse contexts within which researchers and practitioners engage with communities. These commonalities emerge whenever the authors of this book come together to discuss the uniqueness of their individual community engagement models and methods. The individual narratives differ from each other, as they have been influenced by each researcher's/author's discipline, his/her intellectual position in the field, and the nature, scale and location of projects. Irrespective of these differences, some common ground emerges in the way community engagement models and tools have been used.

How to **engage** communities? What is **empowerment**? To what extent should the project process be **participatory**? How is an **outsider-insider** relationship handled? How do researchers negotiate with the **hegemony** of western **cultural interpretations**? How are **organizational and contextual** influences handled in a project? What **leadership** demands do such projects place on researchers? What is **capacity building**? What are **creative leaders** and **creative communities**? How does the researcher journey from their **studio to the situation**?

These questions have been teased out in the individual narratives of the previous nine chapters, and are projected in this chapter through a visual commentary in the following pages. While viewing these visual narratives, readers should be able to situate themselves according to their discipline area/location/nature of project/framework of enquiry, and then compare and contrast the models and methods used. In this way, readers can construct their own personal set of tools for projects that work within communities. They will be able to reflect on assumptions underlying the actions that have been made in their own projects and consider new ways of looking at problems and solving them.

Each narration has three components:

1. A diagram that summarizes the project. Tools that have been used for engaging with the community are listed. The specific tools selected reflect each researcher's intellectual position and context. Summaries are given of the project's nature (teaching and learning, gender specific, skills development, health management and so on), its location (whether in developing/developed/rural/urban context etc.), its scale (dwelling, neighbourhood, settlement etc.) and the discipline that contextualizes it. Capacity building outcomes—which aspect and/or for whom and/or in what way?—are also listed in the same diagram.
2. The second component is the key terminology that defines the unique contributions—the strategies and techniques—that the project/researcher has made to the field of enquiry (community capacity building through community engagement).
3. The last component is a series of images that give a glimpse of the process in which the project materialized, the tools that were used and the people who were involved.

Commencing with Chap. 2, a **collaborative and ethnographic model** becomes the lynchpin around which a place-based architectural project transforms into a product which is developed **by the community and for the community,** using local tools, techniques and skills. Moving on, using a participatory model of engaging communities, Chap. 3 arrives at an understanding that empowerment is about having power **to act with others** in order **to bring change**. Should this participatory model be evaluated by the extent and quality of participation? Chapter 4 dwells on this aspect. It emphasizes that a less inclusive formal participatory process but with a high level of **influence in decision-making** results in a much more effective project because it engenders **strong social capital**. Bringing back the discussion to researchers' and practitioners' cultural and geographical background and their roles as 'outsiders', Chap. 5 proposes a new paradigm of **hybrid practice**. Hybrid practitioners challenge commodified project outcomes that result from a hegemony of western cultural perceptions. The issues of contextual influences and organizational impacts on community projects are discussed in Chap. 6, which suggests a **participatory appraisal framework** to be used by researchers and practitioners prior to commencing any project. Leadership of projects of such a complex nature requires special skills. Chapter 7 describes how **bio mimicry** can be used to tap into nature's resources for learning skills of **creative leadership** with innovative results. Capacity is built, but for whom? Chapter 8 presents a model that provides a **win-win situation for all involved**. Keeping a similar tone, Chap. 9 establishes a **relationship between physical and social space**. Activating the physical space results in an enhanced awareness and learning for all concerned. Chapter 10 returns the discussion to its starting point—an ethnographic engagement. The journey from **studio to situation** is essential, and this creates a demand for researchers, artists, designers, practitioners to involve the locals in a creative process. Through confidence building, transparency and inclusion, barriers are broken. **Creative community is born**.

PHOTOGRAPHS: REENA TIWARI & YATIN PANDYA

DISCIPLINE ARCHITECTURE

NATURE PHYSICAL DESIGN INTERVENTION

THEORETICAL FRAMEWORK
- COLLABORATIVE AND ETHNOGRAPHIC RESEARCH,
- THEORIES OF CO-DESIGN.

LOCATION
- DEVELOPING COUNTRY (INDIA)
- URBAN
- LOWER SOCIO-ECONOMIC NEIGHBOURHOOD

SCALE
- BUILDING
- INTERIOR

CAPACITY BUILDING
- NGO
- COMMUNITY
- LOCAL PRACTITIONERS
- ARCHITECTURE STUDENTS
- RESEARCHERS

TOOLS
- RELATIONSHIP BUILDING WITH STAKEHOLDERS
- OBSERVATION
- PARTICIPANT OBSERVATION
- SCENARIO BUILDING
- CO-DESIGN WORKSHOPS
- INFORMAL CHATS
- UNSTRUCTURED INTERVIEWS
- CO-CONSTRUCTION
- AWARENESS PROGRAMS
- COMMUNITY EVENTS

FIGURE 11.1 MODEL BASED ON CHAPTER 2: 'AN ETHNOGRAPHIC AND COLLABORATIVE MODEL OF INQUIRY: ACTIVITY CENTRE PROJECT IN INDIA' BY REENA TIWARI & YATIN PANDYA

USING LOCAL TOOLS, TECHNIQUES AND SKILLS • URBAN CRAFTING: COLLABORATIVE PLANNING AND ETHNOGRAPHY • BY T

PHOTOGRAPHS: IRENE BASIMIKE, USED WITH PERMISSION

DISCIPLINE
- INTERNATIONAL & GLOBAL HEALTH
- INTERNATIONAL DEVELOPMENT
- RURAL AND MATERNAL HEALTH

NATURE
COMMUNITY HEALTH: HEALTH NEEDS IN RESPECT TO MATERNAL HEALTH

THEORETICAL FRAMEWORK
- PARTICIPATORY AND EMPOWERMENT RESEARCH
- COMMUNITY BASED CAPACITY BUILDING

LOCATION
DEVELOPING COUNTRY (UGANDA) RURAL COMMUNITIES (VILLAGES NEAR JINJA & BUIKWE, IN EASTERN UGANDA)

SCALE
- RURAL
- PERI-URBAN
- A GROUP OF SEVERAL VILLAGES
- MALE AND FEMALE PARTICIPANTS

CAPACITY BUILDING
- COMMUNITY ENGAGEMENT & EMPOWERMENT
- AWARENESS RAISING OF COMMUNITY MATERNAL HEALTH ISSUES
- ENCOURAGING COMMUNITIES TO REALISE THEIR OWN POTENTIAL IN CAPACITY BUILDING
- CAPACITY BUILDING AND COMMUNITY DEVELOPMENT: A COMMUNITY DIALOGUE ON EQUALITY

TOOLS

PARTICIPATORY EMPOWERMENT APPROACHES:
- FOCUS GROUPS
- INTERVIEWS WITH KEY INFORMANTS
- COMMUNITY DIALOGUE
- USE OF CULTURAL INTERPRETERS

FIGURE 11.2 MODEL BASED ON CHAPTER 3: 'CAPACITY BUILDING AND COMMUNITY DEVELOPMENT: A COMMUNITY DIALOGUE ON EQUALITY IN RURAL UGANDA', BY DEBBIE SINGH, MAY LAMPLE, MARK JONES & JAYA EARNEST

MAKING • POWER SHARING • EMPOWERMENT: DEVELOPING THE POWER TO ACT WITH OTHERS TO CREATE CHANGE • PARTI(

IMAGES: MAIN C. MENDOZA-ARROYO & D. LLADÓ-PORTA & INSET C. MENDOZA-ARROYO & ESARQ-UIC FINAL DEGREE STUDENTS 2012-13, USED WITH PERMISSION

NATURE
- NEIGHBORHOOD REGENERATION ASSESSMENT
- PARTICIPATORY TOOLS
- NEIGHBORHOOD REGENERATION ASSESSMENT ACCORDING TO PHYSICAL AND SOCIAL OUPUTS

DISCIPLINE
- URBAN PLANNING
- URBAN DESIGN
- SOCIAL SCIENCES

THEORETICAL FRAMEWORK
- DEMOCRATIC GOVERNANCE
- PUBLIC/PRIVATE PARTICIPATION
- PARTICIPATORY AND INCLUSIVE DECISION MAKING

LOCATION
DEVELOPED COUNTRY: BARCELONA, SPAIN

• INCLUSIVE DECISION MAKING IS THE KEY • LESS INCLUSIVE FORMAL PARTICIPATORY PROCESS BUT MORE INCLUSIVE

FIGURE 11.3 MODEL BASED ON CHAPTER 4: 'URBAN NEIGHBOURHOOD REGENERATION AND COMMUNITY PARTICIPATION: AN UNRESOLVED ISSUE IN THE BARCELONA EXPERIENCE', BY CARMEN MENDOZA-ARROYO & PERE VALL-CASAS

PHOTOGRAPHS: DAMIAN MADIGAN & DAVID MORRIS

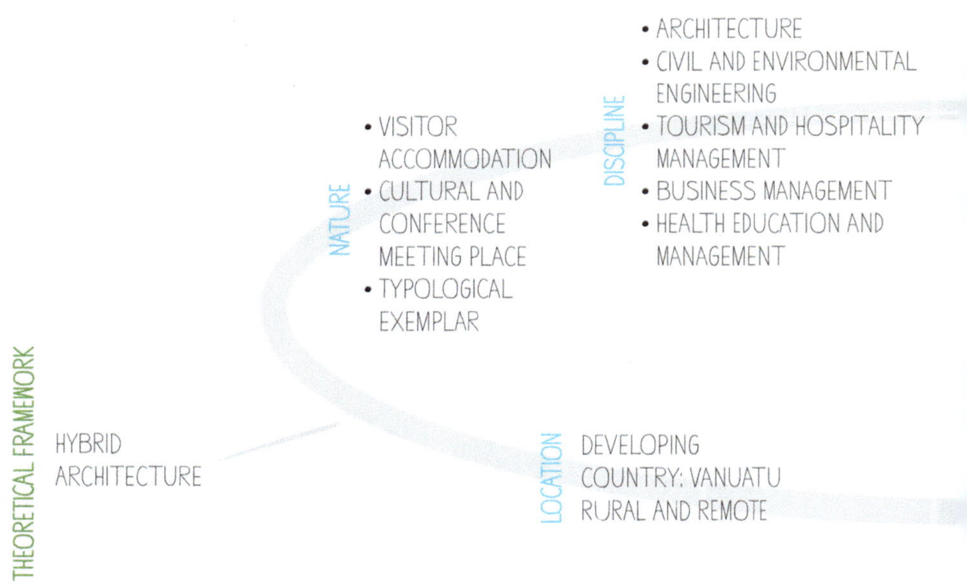

• HYBRID PRACTICE • RELINQUISHING CONTROL • DELIBERATED POSITION • WITH HUMILITY AND EMPATHY • HYBRID

SCALE
- DWELLING
- VILLAGE
- ISLAND

TOOLS
- EXPERIENTAL IMMERSION INTO CONTEXT
- COMMUNITY CONSULTATION
- SITE SURVEY
- ARCHITECTURAL AND ENGINEERING DESIGN AND COMMUNICATION
- PUBLIC EXHIBITION.

CAPACITY BUILDING
- CULTURAL SUSTAINABILITY
- CROSS-CULTURAL AWARENESS
- ECONOMIC INDEPENDENCE
- SOCIO/POLITICAL ENGAGEMENT AND COHESION: AT A UNIFYING COMMUNITY/ISLAND LEVEL AND AT A GLOBAL/VISITOR LEVEL
- ENVIRONMENTAL BUILDING SYSTEMS EDUCATION

FIGURE 11.4 MODEL BASED ON CHAPTER 5: 'THE PORT RESOLUTION PROJECT: DEVELOPING COMMUNITYBUILT AND MANAGED VISITOR ACCOMMODATIONTYPOLOGIES', BY DAMIAN MADIGAN & DAVID MORRIS

PHOTOGRAPHS: CLANCY READ, JAYA EARNEST, MOHAMMED ALI & VEENA POONACHA

NATURE

COMMUNITY HEALTH: ASSESSING, PRIORITISING AND ADDRESSING COMMUNITY HEALTH NEEDS

DISCIPLINE

- PUBLIC / INTERNATIONAL HEALTH
- INTERNATIONAL DEVELOPMENT
- RURAL AND INDIGENOUS HEALTH

THEORETICAL FRAMEWORK

- PARTICIPATORY ACTION RESEARCH (MODIFIED TO CREATE A MORE PRACTICAL APPROACH TO THE METHODOLOGY)
- MULTI-DISCIPLINARY

LOCATION

- DEVELOPING COUNTRY (INDIA)
- RURAL COMMUNITIES (VILLAGES IN GUJARAT, WESTERN INDIA)
- SCHEDULED TRIBE (ST) LOCALITIES (INDIGENOUS, DISADVANTAGED PEOPLES)

• RAPID PARTICIPATORY APPRAISAL • KNOWLEDGE TO ACTION TRANSITION; ACTION-RESEARCH USING LOCAL TOOLS

SCALE

- REGIONAL (TALUKA LEVEL) NB. A TALUKA IS A SUBDIVISION OF A DISTRICT; A GROUP OF SEVERAL VILLAGES ORGANIZED FOR REVENUE PURPOSES
- VILLAGE COMMUNITY (5 VILLAGES IN TOTAL)

CAPACITY BUILDING

- LEARNING
- SOCIAL ENGAGEMENT & AWARENESS RAISING OF COMMUNITY HEALTH ISSUES
- ENCOURAGING COMMUNITIES TO REALISE THEIR OWN POTENTIAL IN ADDRESSING THEIR COMMUNITY NEEDS

TOOLS

- PARTICIPATORY ACTION RESEARCH METHODOLOGY
- PARTICIPATORY APPROACHES (RAPID PARTICIPATORY APPRAISAL [RPA] & COMMUNITY-BASED PARTICIPATORY RESEARCH [CBPR])
- RPA OBSERVATION
- FOCUS GROUPS AND INTERVIEWS WITH KEY INFORMANTS
- SECONDARY DATA/DOCUMENTS
- CBPR: COMMUNITY MEETINGS

FIGURE 11.5 MODEL BASED ON CHAPTER 6: 'APPLYING A PRACTICAL, PARTICIPATORY ACTION RESEARCH FRAMEWORK FOR PRODUCING KNOWLEDGE, ACTION AND CHANGE IN COMMUNITIES: A HEALTH CASE STUDY FROM GUJARAT, WESTERN INDIA', BY CLANCY READ, JAYA EARNEST, MOHAMMED ALI & VEENA POONACHA

TECHNIQUES AND SKILLS • RAPID PARTICIPATORY APPRAISAL KNOWLEDGE TO ACTION TRANSITION; ACTION-RESEARCH •

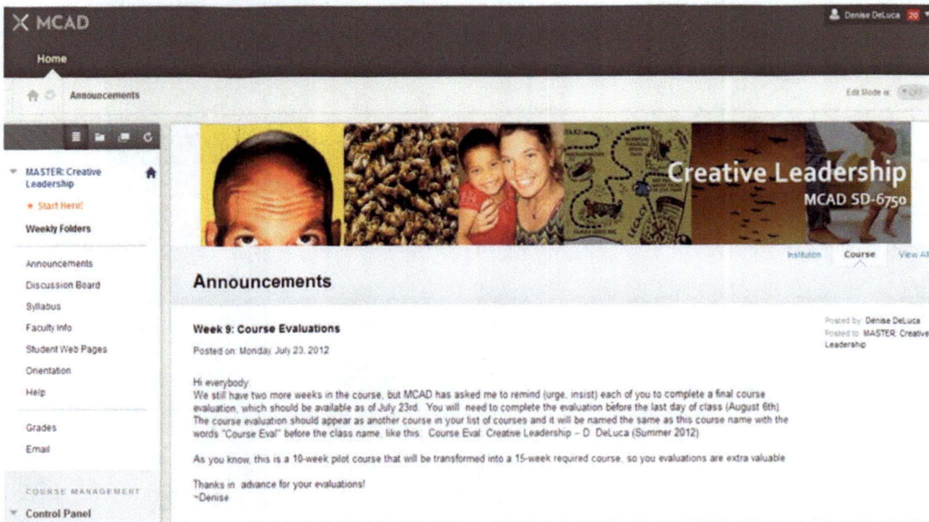

IMAGES: DENISE DE LUCA, SHANNON RAHKOLA & NOBLE CUMMING, USED WITH PERMISSION

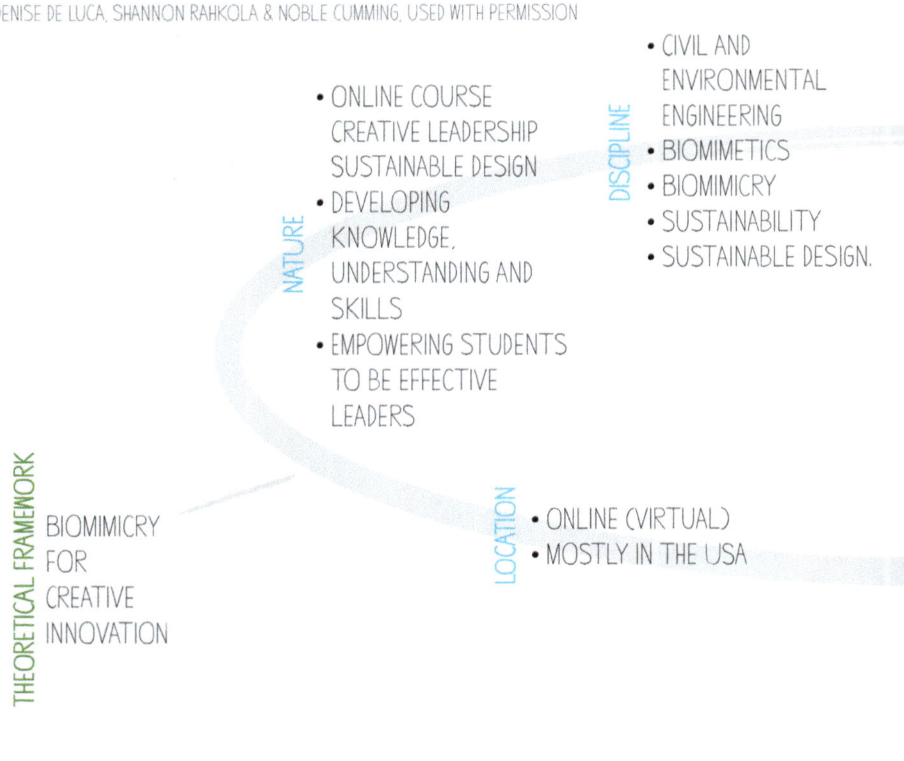

• CREATIVE LEADERSHIP RADICAL INNOVATION USING BIO-MIMICRY USING LOCAL TOOLS, TECHNIQUES AND SKILLS •

SCALE
- VIRTUAL CLASSROOM
- MASTER'S DEGREE PROGRAM

CAPACITY BUILDING
- INTELLECTUAL (KNOWLEDGE)
- EMOTIONAL (UNDERSTANDING)
- PHYSICAL (SKILLS)

TOOLS
- COURSE DEVELOPMENT AND TEACHING
- POST-COURSE FEEDBACK

FIGURE 11.6 MODEL BASED ON CHAPTER 7: 'INSPIRED BY NATURE: BUILDING COMMUNITY CAPACITY THROUGH CREATIVE LEADERSHIP', BY DENISE DE LUCA

CREATIVE LEADERSHIP RADICAL INNOVATION USING BIO-MIMICRY USING LOCAL TOOLS, TECHNIQUES AND SKILLS • CRI

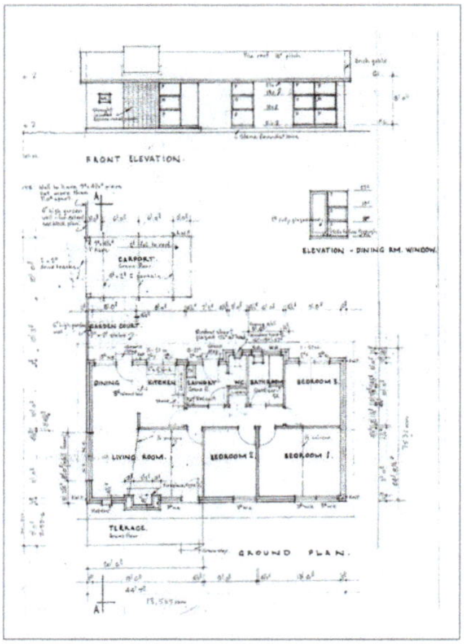

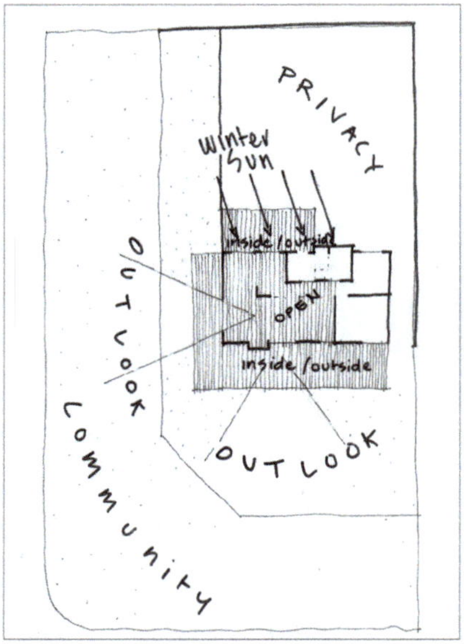

DRAWINGS: SARAH MCGANN PHOTOGRAPH: DAVID ROBINSON, USED WITH PERMISSION

DISCIPLINE
- ARCHITECTURE
- DESIGN

NATURE
PRODUCTION AND TEXT

THEORETICAL FRAMEWORK
RESEARCH AND PRAXIS NEXUS

LOCATION
- DEVELOPED COUNTRY (AUSTRALIA)
- SUBURBAN

• COMMUNITY BASED PRO-BONO DESIGN • COLLABORATIVE WORK • BUILDING CAPACITY FOR ALL • RESEARCH •

SCALE DWELLING

CAPACITY BUILDING
- COMMUNITY NEED: PROBONO PROJECT (ECONOMIC AND PHYSICAL)
- RESEARCH AND CREATIVE PRACTICE GAP
- HOPE
- FUTURE VISION/ ACTION

TOOLS
- DESIGN THINKING AND DESIGN WRITING
- CREATIVE PRACTICE AND VISUAL ESSAY
- RESEARCH SYSTEMS

FIGURE 11.7 MODEL BASED ON CHAPTER 8: 'CONNECTIONS: ACADEMICS, ARCHITECTS AND COMMUNITY PRO-BONO PROJECTS', BY SARAH MCGANN & BARBARA MILECH

PRAXIS • COMMUNITY BASED PRO-BONO DESIGN • COLLABORATIVE WORK • BUILDING CAPACITY FOR ALL • RESEAR

PHOTOGRAPHS: LEFT ANA CALIC & RIGHT LARA MACKINTOSH, USED WITH PERMISSION

• STREET ACTIVATION – A CATALYST FOR CHANGE • EXHIBITION • ENGAGEMENT • ENHANCED AWARENESS • STREET A

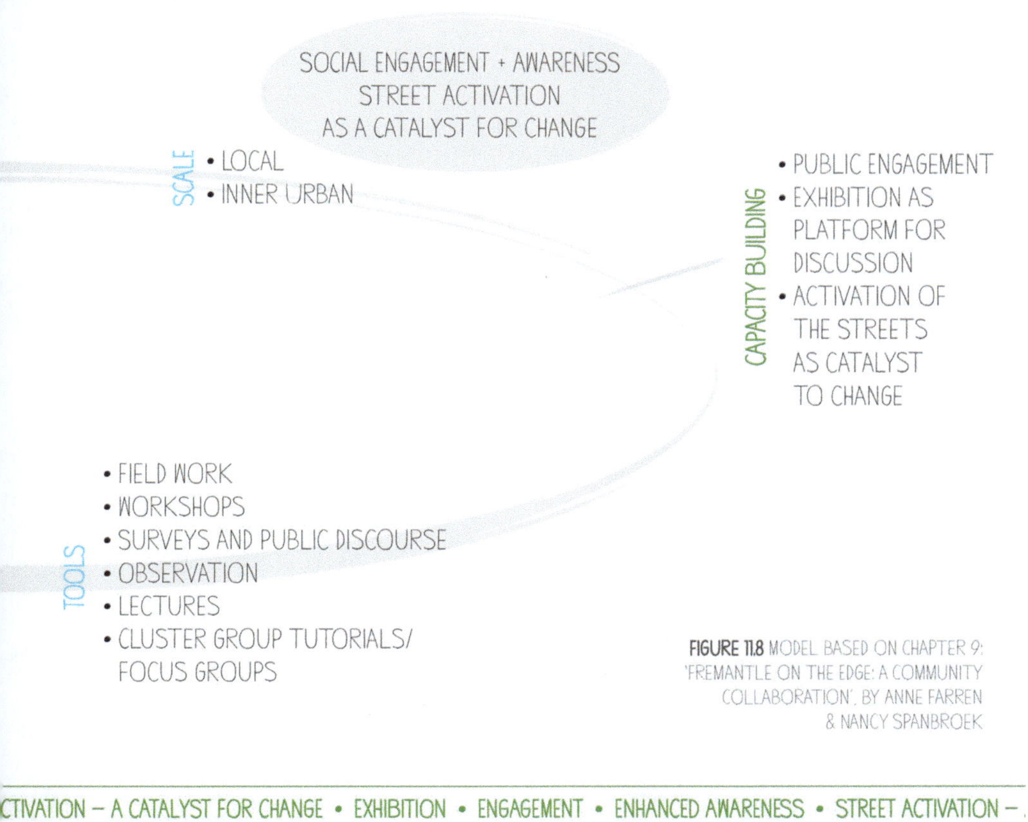

FIGURE 11.8 MODEL BASED ON CHAPTER 9: 'FREMANTLE ON THE EDGE: A COMMUNITY COLLABORATION', BY ANNE FARREN & NANCY SPANBROEK

PHOTOGRAPHS: SALVATORE DI MAURO

• CREATIVE COMMUNITIES • CREATIVE PRACTICIONER • STUDIO TO SITUATION • CREATIVE COMMUNITIES • CREATIVE PRAC

SCALE

- LOCAL RURAL AND URBAN COMMUNITIES
- INCLUSIVITY OVER EXCLUSIVITY

CAPACITY BUILDING

- UNDERSTANDING THE CREATIVE PROCESS
- CONFIDENCE BUILDING THROUGH TRANSPARENCY
- EQUITY AND INCLUSION
- BREAKING DOWN BARRIERS AND NEGATING DIFFERENCE

TOOLS

- PRELIMINARY ARCHIVAL AND LOCAL RESEARCH
- EXHIBITION SHOWCASE/RELATIONAL AESTHETICS
- STUDIO TO SITUATION
- DISPLAY
- SOUNDING BOARD AND THE ACT OF DISCLOSURE
- WORKABLE TEMPLATES
- PROGRESSIVE EVALUATIONS
- HANDS ON ACTIVITY

FIGURE 11.9 MODEL BASED ON CHAPTER 10: 'TRANSPARENCY AND INTERDEPENDENCE', BY SALVATORE DI MAURO

TICIONER • STUDIO TO SITUATION • CREATIVE COMMUNITIES • CREATIVE PRACTICIONER • STUDIO TO SITUATION • CREATI

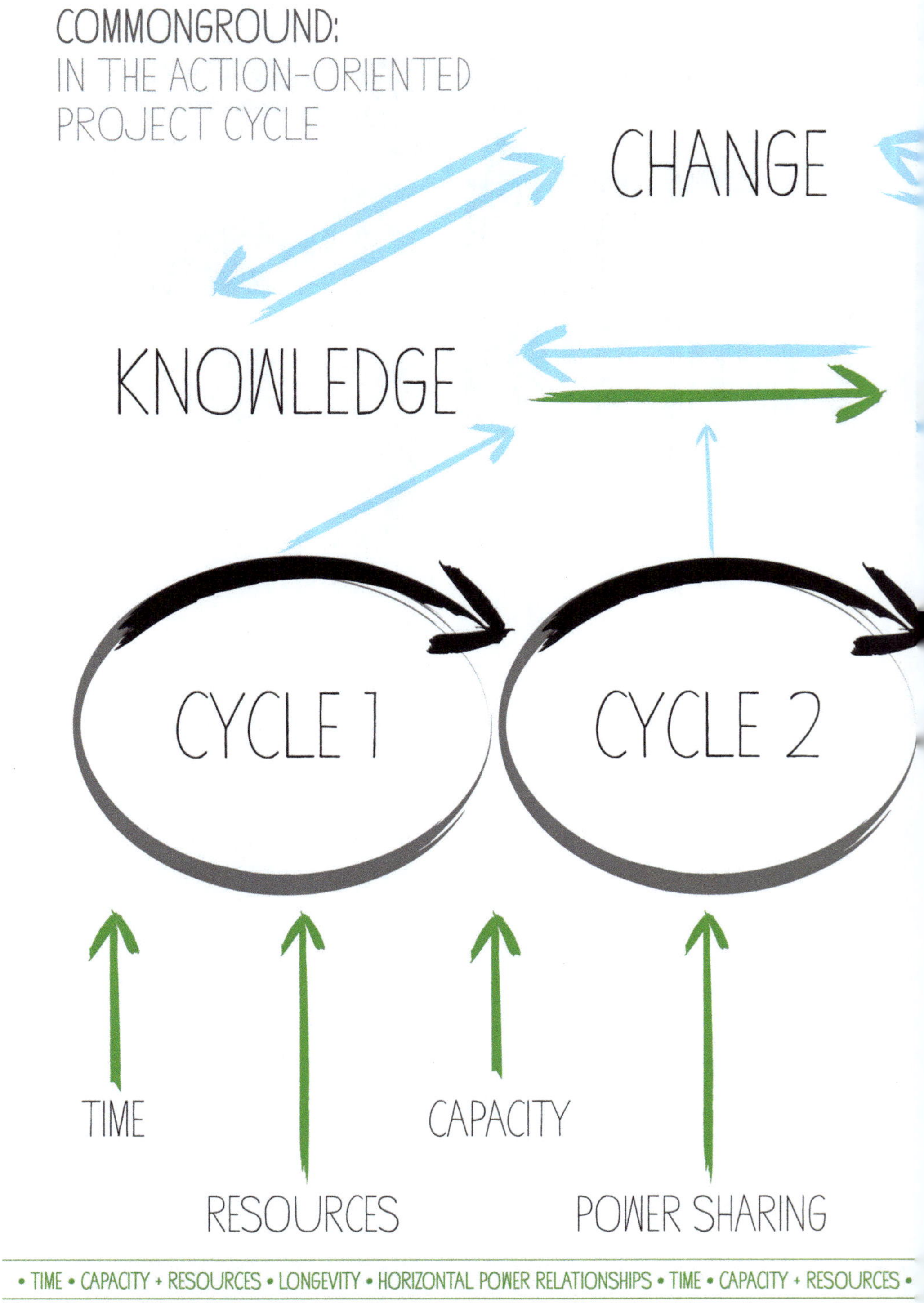
COMMONGROUND:
IN THE ACTION-ORIENTED PROJECT CYCLE
CHANGE
KNOWLEDGE
CYCLE 1
CYCLE 2
TIME
CAPACITY
RESOURCES
POWER SHARING
• TIME • CAPACITY + RESOURCES • LONGEVITY • HORIZONTAL POWER RELATIONSHIPS • TIME • CAPACITY + RESOURCES •

FIGURE 11.10 COMMONGROUND IN THE ACTION-ORIENTED PROJECT CYCLE (TOP TRIANGLE: THE KNOWLEDGE, ACTION, CHANGE RELATIONSHIP, IS BASED ON READ 2012, P. 306)

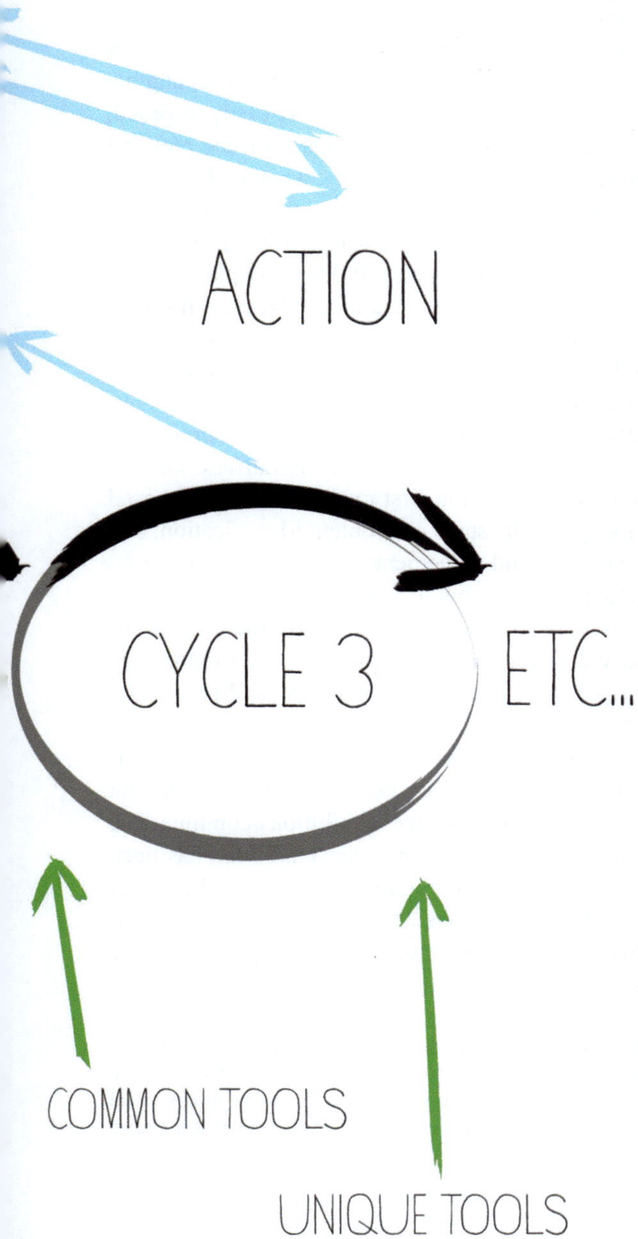

TIME: CHALLENGES ABILITY TO TRANSLATE KNOWLEDGE INTO ACTION • TIME OF RESEARCHERS RELATED TO ACADEMIC CALENDER • TIME OF COMMUNITY MEMBERS & THEIR COMMITMENTS • TIME FRAME OF THE PROJECT AS DEFINED BY THE FUNDING BODY LONGEVITY: AFFECTS CAPACITY BUILDING/EFFECTIVENESS • NUMBER OF ACTION RESEARCH/PROJECT CYCLES • RELATIONSHIP BUILDING • ONGOING CAPACITY BUILDING • ABILITY TO CHART PROGRESS • CAPACITY + RESOURCES: CONSTRICT COMMITMENT TO AN ONGOING PROGRAM, REQUIRED TO BUILD THE CAPACITY OF COMMUNITY AGREED INITIATIVES • OFTEN STALLED AT FIRST CYCLE, AT PROBLEM IDENTIFICATION AND NEXT ACTION • HORIZONTAL POWER RELATIONSHIPS EMPLOYED • THOSE OUTSIDE THE COMMUNITY EMPHASIZE THE NEED TO LEAVE THE BAGGAGE OF POWER TO THE SITUATION • A TWO (OR MORE) -WAY KNOWLEDGE-SHARING FRAMEWORK • TRUST • COMMON TOOLS USED • WORKSHOPS • FORUMS • ETC... • UNIQUE TOOLS EMERGED • EXHIBITION: COMMUNICATING THE OUTCOMES VISUALLY AIDS COMMUNITY UNDERSTANDING • DIFFERENT METHODS OF COMMUNICATION USED TO TRANSCEND THE COMMUNICATION BARRIERSN PHYSICAL SCALE MODELS OF CONCEPTS • TIME: CHALLENGES ABILITY TO TRANSLATE KNOWLEDGE INTO ACTION • TIME OF RESEARCHERS RELATED TO ACADEMIC CALENDER • TIME OF COMMUNITY MEMBERS & THEIR COMMITMENTS • TIME FRAME OF THE PROJECT AS DEFINED BY THE FUNDING BODY LONGEVITY: AFFECTS CAPACITY BUILD-

LONGEVITY • HORIZONTAL POWER RELATIONSHIPS • TIME • CAPACITY + RESOURCES • LONGEVITY • HORIZONTAL POWER REL

Conclusion

The visual narration of the projects and their process has operated as a self-reflective journey for the authors. Authors diagrammed their projects, analysed the process, and reconsidered and questioned their experiences within a broad context of location, scale and discipline that affected each project's capacity building outcomes. The projects that were small scale allowed a quick measure of the project's success. The tangible benefits (for example, of built environment projects) were much more easily recorded, while recording intangible benefits, as in health and educational projects, required sophisticated auditing and interpreting tools. Locational aspects—whether the context was a developing or developed country, or a rural or urban setting—contributed to specific and different barriers and challenges; the next two chapters elaborate on these.

Time was a limiting factor that all the case studies shared. For example, the time limits of researchers working within the academic calendar often placed restrictions on what could be accomplished, as did the time frame of each project itself as defined, for instance, by the funding body. The longevity of a project, constrained by these limits, affects the effectiveness of capacity building.

Finances constrict change agents' ability to commit to an ongoing program, and are usually required to build the capacity of community-based initiatives. For example, a number of the case studies illustrated the first cycle of an action-based cycle, where the various players had got to the stage of problem identification, and developed the relationships around change and agreement about the way forward.

Power relationships were implied in all cases, but in a number of these, researchers, as outsiders to the community, emphasized the need to respect the capacity of the 'insiders' in the situation to make decisions. For example, a two- (or more) way knowledge-sharing framework was set up instead of a commonly held perception that the 'outsider' carries expert knowledge.

Common tools were used in many of the examples, such as workshops, forums for all. Unique tools also emerged in some of the case studies, ones which were not commonly used across the other examples. For instance, exhibitions communicate the outcomes visually in order that the community can understand what has been envisioned and achieved. Different methods of communication were used in order to communicate with a variety of different members and transcend the communication barriers. For example, visual means of communication like physical scale models, installations and drawings were used to overcome language barriers.

Acknowledgements We would like to acknowledge the contributions of contributing authors to Part I for their diagrams, and the contributions of Bob Pokrant (anthropology and social sciences), and Mohammed Ali (health sciences) for providing the different disciplinary perspectives in this visual essay. And also Allison Crimp for graphics, and Roblyn Creagh for preliminary work on visual essays.

References

Read, C. (2012). Applying a participatory action research model to assess and address community health concerns among tribal communities in Gujarat, Western India: The potential and challenges of participatory approaches (Doctoral thesis, Curtin University). Australian Digital Thesis Website. http://espace.library.curtin.edu.au:80/R?func=dbin-jump-full&local_base=gen01-era02&object_id=188776. Accessed: 27 Oct 2013.

Chapter 12
Joining Communities: A Role for Reflection

Dianne Smith

Abstract In this chapter, I aim to provide newcomers to community-based projects with a summary of clues or hints about how they may approach such projects. Many suggestions relate to being aware of how a project may unfold regardless of its scale, location or complexity. The building of a solid, informed and honest team is of importance. The checklist included is built from personal reflections, drawn from each community-based project I have undertaken, with the aim of building greater sensitivity and awareness of community needs and the circumstances that surround them.

Keywords Reflection · Reflective practice · Community

Introduction

'Out of the pot and into the fire' is a saying that aptly describes how we may feel increasingly challenged personally and professionally when we chose to move into new areas, however complex and challenging our professional life already is. What we have taken as known will need to be reconsidered. The interface between university and practice is characterized by this disjunction (Schmidt 2000). For a beginning professional—scientist, doctor, architect, designer, sociologist, musician, and so on—employment brings the reality of working within actual situations with identifiable stakeholders. Although shifts in late twentieth century professional education have aimed to provide embedded theory and experiential learning situations to reflect what occurs in professional practice, with employment comes the reality of being responsible for the tasks at hand and engaging in actual professional projects which have tangible consequences.

Similarly, for those entering community work, either as an academic or as a discipline expert, there is a need to consider what it is to be a novice in this new type

D. Smith (✉)
Department of Architecture and Interior Architecture, Curtin University, Perth, Australia
e-mail: dianne.smith@curtin.edu.au

R. Tiwari et al. (eds.), *M² Models and Methodologies for Community Engagement,*
DOI 10.1007/978-981-4585-11-8_12, © Springer Science+Business Media Singapore 2014

of domain. Therefore, in this discussion I will highlight the role of critical reflection and action research (the cycle of doing, reflecting, proposing action, doing) in the professional development of a community engagement practitioner. Moon (1999) discusses the theory of reflection and the variations in understanding of the term. She identifies reflections in five ways: general interpretations, that is, reflection as part of the process of learning; when considering something in more detail, toward a useful outcome; during mental processing where purpose or outcome is unclear (beyond recalling); as a tool for professional practice; and/or learning from experience. Moon states: 'reflection itself is a mental process that seems to be related to thinking and learning', and differences in types of reflection lie in the way that the process is 'used, applied and guided' (1999, p. 5).

Through a series of reflections upon my own encounters with community-based projects, I will highlight a number of barriers and ways to reconsider and/or actively address them. In order to undertake this task, I will first revisit the concepts raised by Donald Schön and other researchers in their discussion of reflective practice as a context for community-based project work as well as for the practitioner's personal development.

The Reflective Practitioner

The danger for professionals who are discipline experts when they shift from their core context to a community engagement project is that the need for new knowledge and skills may not be immediately evident or recognized. Just as a world-leading expert in a particular field (such as law, accountancy, art, or anthropology) may not recognize that they have no guarantee of being an excellent lecturer or academic when they join a university, a shift from the academy to a community-based project work is not guaranteed success. The new role, in both examples, requires a different or extended field of expertise.

Change requires reconsideration of one's personal practices and an assessment of what new skill sets, knowledge, and actions are necessary. How will these potential acquisitions facilitate the development of this emergent dimension to their career and their life experience? Self-awareness and a willingness to evolve are attributes which Schön, in his book *The Reflective Practitioner* (1983) and his followers (e.g. Billett 2010), have alluded to as being essential for professional practitioners.

The key to success is the development of a repertoire (Schmidt 2000); that is, the life and professional experiences that create a wealth of knowledge and which can be drawn upon during projects, particularly when confronted by the complexity of new situations with their embedded issues and problems. Schön refers to the duality of high theory and the 'lowly swamp' of real life experiences for professional practitioners; he states that in order to apply the theoretical principles to everyday projects, we require the rigorous testing of theory while ensuring its value once it is applied to 'the swamp' (Schön 1983, as cited in Bertolini et al. 2010; Schön 2001). Schön states it is reflection that enables such synergies to occur; and therefore, by

Table 12.1 Reflective-practitioner modes of practice

Professional practitioners	Novice community engagement researcher
Developed level of practice expertise	Acquisition of experience
Skill set & knowledge repertoire	Learning through doing
Reflection	Reflection
Reflect-in-practice	Reflect-in-action

extrapolation, reflection is highly relevant to the move from graduation to practice, the shift from practice to being a lecturer, or the entrance by an expert to novel community engagement situations. Therefore, it is the nature of reflection that I will now address.

Many examples are provided in the literature. However, the distinction between reflection-in-action and reflection-on-action provides a useful point when we consider how working in community projects can be improved over time and through sharing our experiences. Reflection-in-action occurs while we are participating in the job-at-hand (Gero and Kannengiesser 2008). We thereby come to consider what we are doing, how, and why, as well as the consequences if the predicted or expected norms are not eventuating. We are required to deal with a particular issue or situation at a specific time and place. What occurs is what Schmidt calls 'thought experiments'; these reflections and explorations yield a 'feel of the situation' and refinement of our 'knowledge in action' (Schmidt 2000, p. 270). However, it is also necessary to avoid professional complacency. That is, as Schmidt warns, although a repertoire is developed, we must always be on the look out for the special cases that may fall outside our suite of tools or our manner of looking at things (Schmidt 2000). Table 12.1 presents the shift from competent reflective practitioner projects to novel situations, whereupon we need to draw on our repertoire as well as learning through reflective practice about the uniqueness of the situation.

Each community project is unique, particularly in contexts where content, culture and people are unfamiliar to us. Such surprises stimulate reflection-in-action (Schmidt 2000), which, in turn, facilitates the development of ways of dealing with whatever arises by drawing on past knowledge combined with that arising from the current situation. Reflection is a means to knowledge and understanding (Gero and Kannengiesser 2008).

In contrast, the act of reflection-on-practice results in seeing what has been done before and what the impact has been. And when comparing different situations, we can compare the current context and issues with those in our past experiences. We can 'immerse [our]selves in the situation and make deliberative moves with the intent to change things and also playful exploratory probes just to get a feel for things. Both designers and practitioners go through stages of reflection, appreciation, and action' (Schmidt 2000, p. 272).

'Reflection allows practitioners to change the way they go about solving problems' (Gero and Kannengiesser 2008, p. 1). In association, as the sociologist Erwin Goffman stated, it is through social interactions that we develop understandings and sense of self (Goffman 1959). The reactions of other individuals and collectives can reinforce a person's identity and constructs of appropriateness and belonging. In the context of community engagement projects, the participants and setting can serve a similar role—enabling insights into the way we practice as well as about the phenomena of working on projects that involve multiple players. Each player brings something to the project, yet each of these can reinforce or disturb the others' premises and modes of operating.

What is evident is that community engagement as a practice can model professional practice, and thereby, both benefit through reflection; and in association, the application of critical action research enables a means to improve one's own practice by integrating the technical aspects with the larger social and moral issues involved (Capobianco and Feldman 2010, cited in Hagevik 2012). Building understanding and navigating future scenarios through immersion in the project provides a sense of what is happening over and above the describable facets of the situation; this is a phenomenon Bourdieu termed habitus (Bourdieu 1986). A person's habitus influences the way they practice—that is, in this case, how they conceptualize and engage in community work.

By looking back over my experiences of community projects and examples of community engagement and reflecting upon what was learnt, it is possible to construct a 'hint-list' for new players. In this way, cautionary notes, barriers and opportunities are revealed that may inform the planning of my own new projects, and also inform others in regard to their projects.

Reflective Case Studies

In this chapter I will relate my experiences of projects that I have participated in over the past twenty years of academic practice. It needs to be stated that my recollections and summations may not reflect those of other collaborators. However, it is not my intent to edit the cases cited to fit others' impressions. Rather, the aim is to develop insights in order to inform others about community engagement across multiple sites and situations, as a catalyst for their own reflections upon their experiences, and for evaluation of any plans for future work. The aim is to identify the diverse aspects that need to be considered when asking: what are the significant aspects that can influence the success or failure of community engagement projects? The overall findings may be either a unique attribute or a more typical project focus (or maybe both) in such community-based interactions; regardless, they are stimuli for reflection and critique.

Five contrasting projects are used as the basis of this reflection-on-action. These are listed in Table 12.2.

Each project will be described briefly, prior to a summary of lessons learnt and the implications of such insights for future projects. The names of these places and organizations are irrelevant to these tales, and therefore, the nature of the place and project are prioritized over and above the where and who. This serves to maintain anonymity of the players referred to in regard to the issues discussed. This overview

Table 12.2 Selected community engagement projects

	Action	Location	Context
1.	Mainstreet project	Small town	QLD coastal rural
2.	Refurbishment	Service-provider facility	QLD urban inner city
3.	Design scheme	Community centre	WA regional centre
4.	Occupancy survey	Small village	International rural
5.	Design scheme	Education facility	QLD remote rural

is followed by a summary of suggestions for newcomers to community engagement, built from these reflections.

Project One: Townscapes

Description

The setting is a small rural town in south-east Queensland that attracts an influx of seasonal workers to assist with the annual harvests. It is located near the State's main north-south thoroughfare, close to a beautiful coastline and national park. Its aim, over 20 years ago, was to revamp its main street and profile in order to attract travellers to stop or visit the town as a destination. Our team's involvement in the community project came about through a connection between the townspeople and a colleague who is a designer and artist. The initial contact led to the team of students and staff from two university design programs becoming involved to undertake contextual studies and survey the locals to assess the character of the town and its aspirations. These initial activities were the forerunners to a broader and professionally-led team engaged in community workshops, consultation procedures, an exhibition of the findings, and a report to the council, as well as ongoing media releases to inform the public. The project lasted for over two years, so that links with community members and local government representatives and business owners had time to develop.

Pivotal Point

On reflection, the pivotal learning point was a discussion concerning the future of the project. This was part of a community workshop where two conflicting concepts of the project's process were disclosed. This was a valuable lesson for naive players. On one hand, the project group envisaged that the general public at the workshop owned the project, and we felt that many participants trusted us to drive an inclusive agenda toward change for their town. In this belief, the project collaborators wished to protect the open nature of the consultative and collaborative process, as well as the need for discussion and engagement as an endpoint for the project that was being created; in short, a strategy and design. On the other hand, the council disclosed their understanding of the project group as facilitators who, once data was collected, would move on, so that the council would drive the future project development. At the workshop, these two parallel yet contrasting positions became clear. Ironically, this revelation occurred because a question the project team asked (as part of the inclusive process) of the wider community was 'where to go from here?'. This brought an answer from a discrete sub-group, the local government representative: he would take over the process. This decision, however justified it may have seemed to the representative, disempowered the team, and we felt that many of the community who might not normally attend such activities or partake in the conversations associated with them were also disempowered. We felt we had let down these community groups by allowing one stakeholder group to take over, during a forum where all, under the team's guidance, were supposed to be equal players.

Insights

- Friendships are built with community members, but care needs to be taken in terms of expectations and the equality of players.
- Within all communities are factions, some operating with goodwill toward others, others generating tensions and constrained by implicit power relationships. The ability to recognize the level of flexibility for a group, and the means and sensitivity required to facilitate the acquisition of greater understanding between groups, are valuable attributes in community engagement projects.
- Access is fragile. Even with an invitation, members of a community engagement team are not locals, although friendships grow over time and with increasing familiarity. Respect for this fragility and the real and potential consequences of being an *insider*, or being locked on the *outside*, needs to be considered at the project's outset.

Project Two: Sub-Cultural Groups

Description

Project two's brief for an inner city youth centre was to assist with the upgrade of their small two storey inner city building by creating a friendlier, safer drop-in centre for youth at risk. In addition, it was to be an efficient and safe work environment for administration staff and case workers. Open during the day, the centre offered young people a place of support as well as friendship, activities, and services, including computer searches, access to accommodation sources, medical personnel and councillors. The team's goal was to provide design suggestions that could form the basis for funding applications to enable the upgrade.

This project contrasts with the previous one in almost all dimensions. Our team only involved four people—two students and two staff, who worked with the providers to identify their needs, and with the youth to gain insight into their aspirations as well as their local skill sets and knowledge. However, there was the additional consideration of client, staff and personal safety, which raised questions regarding who needs to be protected from whom. There was also the potential opportunity or need for advocacy that is often associated with community-based research. Youth-at-risk (due to homelessness, domestic violence, broken homes, mental illness, substance abuse, cultural issues) are a group whose voices need support to be heard. The intimate scale of this project gave the people, place and issues an immediacy and importance that can sometimes be lost or diluted in larger and logistically demanding projects.

Pivotal Point

Our value was to bring an outsider's viewpoint to the task, a view that the locals most likely cannot see, because they are immersed in the situation. As close relationships are likely to form, especially on long-term projects or ones of an intimate nature, a balance between being empathetic and informed is needed. There was a need to be aware of the local contextual (social, economic, political) influences, to avoid false expectations that definitely could not be met. For example, this project's role was to attract funding, and there was likely to be an unavoidable delay between the end of our project and the securing of funds to do the upgrade. This time lag can lead to a breakdown in the process and of the connections formed. Key people move on, organizational structures are restructured, and differing government departments or policy makers have conflicting and evolving priorities and willingness to take risks or invest in fringe projects. Therefore, the impact of the inherent delays needs to be considered at the onset of community engagement projects.

Insights

- Going into community implies that the researcher is foreign to this particular situation, regardless of good intentions and preparation. This is amplified where risk and safety are involved. The researchers' emotions need to be recognized; if they are comfortable, there is more opportunity to engage openly with the participants in a meaningful way. To create a zone of comfort, personal or discipline assumptions about clients/users need to be made explicit within the team before embarking on the project, so that shared knowledge can modify any ill-informed positions, and insights and processes can be built that reflect an open and informed understanding of people and place. There is much potential for researchers, consultants and students to come to know different people, and through these connections to understand the project, themselves, and the impact that change may bring.

- Just as community members are vulnerable when change is happening and outside agents are involved, so is the researcher. The scale of a project creates different levels of vulnerability for the researcher because, if small, researchers are in close proximity spatially and functionally. Also, when projects are small and intense, there are advantages and disadvantages of high personal investment (of time, effort and emotion for example) as well as personal connectedness (through developing relations including friendships); and therefore, a researcher's long-term usefulness to a community requires the research team to balance this with what they perceive regarding the immediate situation and events—not just the immediate impact per se. In contrast, larger, complex projects run the risk of the researcher objectifying the issues; that is, their processes may unintentionally dismiss the immediate and/or overall needs and desires of the community members at a local or individual level. The immediate impact, as well as long-term outcomes, may impact on the project's value for the community involved, and this needs to be considered.
- Often, researchers are privileged through their education and access to travel, and indirectly or directly have opportunities to lobby or converse with governments or other power brokers. Therefore, the researchers may be able to give voice for communities directly and/or through the research. However, this role brings with it an important responsibility to understand the insiders' position rather than be driven by the desire to 'fix' what the 'outsider' deems to be incorrect.

Project Three: Diverse Groups

Description

This student-focused project, in a regional rural centre in the Western Australian wheat belt, required us to design a community centre for the township. Key aspects of the town were the migrant populations, indigenous culture and rural life, and demands that exposed the team to a wide range of issues including social justice aspects such as equity and discrimination. Students and staff lived in the town for approximately a week, during which they were briefed, spoke to and observed locals, attended site visits, and set up in a local building to become a project-studio. The concept was to work with community, and to gain their feedback and input over the period. Our team did not speak the breadth of languages spoken by the locals, many of whom were migrants working in the local industries and their family members. Others were residing on properties outside the town, were Indigenous Australians, and others were busy in their businesses or service provision. The limitations on opportunities to engage in deep conversation across sectors of the community meant student-researcher/designers felt their ability to complete the project sensitively and thoroughly was compromised.

Pivotal Point

What was highlighted was the need for continual contact between players during the project's life to ensure that those who facilitate community access are informed; it is important to have a project-contact person for the team and to inform the team and community of what is happening. It is also necessary to gain insight into who the gatekeepers of information are, and what is appropriate in relation to process, engagement, and distribution of ideas. This is particularly important where one of the project's broader objectives is to facilitate input from as many minority groups as possible, in addition to the more prominent subgroups. Further, there is a need to navigate conflicting and sensitive information and opinion (particularly when some groups or individuals seem to have more authority) in order to give all input credence, and then to evaluate the disparate input within the overall context and complexities of the project. Indirectly, non-disclosure by some community members and sectors can place a project team in a difficult position, and their ability to act with authenticity and trust is compromised.

Insights

- Identifying the unknowns and the informal power groups and their associated dynamics early may save time and energy, as well as avoiding a build-up of false expectations due to one's own biases.
- Degrees of genuineness impact on the members of the entire project—community and researchers. When people genuinely wish to contribute and to facilitate change or improvement, there is risk that a lack of truthfulness by some members can undermine meaningful relations and also other outcomes that may have evolved (if the full picture was understood).
- Honesty of communication and agreed rules of interaction, including disclosure, may assist team members to address more complex issues and achieve outcomes. Such guidelines need to be established in the planning phase.
- Having a team which can empathize and communicate with the diverse members of the community can enrich the project.

Project Four: Village Issues

Description

International projects bring additional complexities because of the 'exotic' setting and the researchers' lack of first-hand knowledge. This project is set in a small rural village in Northern India surrounded by farmland. Although the project originated through an NGO who wished to improve the locals' access to education, the overall objective of our initial visits was to survey the village (infrastructure and houses) and its people to identify what the key issues are. The project is cross-disciplinary and, although ongoing, initially involved approximately a week of visits to the site, as well as studio-intensives back at the hotel to process and develop the workz. A community welcome commenced the visit, and a presentation to the villagers at the end spoke of the findings and the emerging plan for the future. Due to the cost of travelling to India, unique constraints were evident; for example, the pressure to justify value for the budget meant the scope of the core project needed to be expanded to include additional outcomes; to put it crudely—more outputs for the dollar invested. The balance between being able to participate logistically (in this case being funded to attend) and the scope and need of the project (reflected in the sub-projects selected) is a fragile one. The value and impact need to be calculated with care and realism in terms of a particular community; this may lead to a decision not to be involved in a project, if needs and impact, when reviewed realistically, actually favour the research project team's objectives over the needs of the community involved.

Pivotal Point

Understanding the influences on interaction and the impact on the project's value is a pivotal aspect of this project. Two initial characteristics stand out for me: firstly, the need to travel to the village every day, with the group arriving and leaving daily, emphasized the external nature—and the potential limitation—of its involvement; and secondly, the inability to speak to the locals because I (and 98 % of the other team members) cannot speak Hindi. This aspect became more evident when returning the following year and meeting people whom I had met and engaged with in activities previously, and yet not being able to engage with them as old acquaintances. This invisible barrier is also potent because we were visiting individual's homes, and walking through their narrow streets within a township that could be circumnavigated by a 30 minute brisk walk.

Another influence on the modes of engagement is the need for mediators. For example, as the NGO had requested our involvement in assessing the potential of education buildings for improvement, the relationship between the university and villagers is always filtered by the NGO's objectives. The evolving relationships between the three or four players is critical (our research team, the NGO, the community leaders, and the individual residents and community groupings). Developing and navigating a process that enables relevant positive change, yet is authentic for the local community members, is maybe the most critical yet difficult task.

Insights

- The scale of the interaction is an important consideration in addition to the scale of the project per se. Levels of intimacy and proximity influence the way the researcher is encountered and therefore perceived, and vice versa.

- The shift from being an outsider to a collaborator needs to be considered and facilitated. Often the logistical considerations may interfere with opportunities for immersion, team building and positive modes of interaction.
- A lead organization (such as an NGO) will create a filtered form of engagement for the other project players. Therefore, when creating teams or partnerships, attention to building open communication, exchange of ideas and frank discussions about processes, may help to mitigate difficulties arising because of the filters people have.
- Agendas outside the project can influence the team members' interpretation of their scope and modes of engagement in an effort to comply, so as not to jeopardize the overall project. Project teams should be aware of 'creep'—a state by which small adjustments may undermine or taint the original or most significant goals of the core project.
- The rationale for being involved needs to be assessed– knowing when to say *yes* and when *no* can be facilitated through reflective practice.

Project Five: Community Purpose

Description

This project was set in a small isolated town located on the western highway through rural remote Eastern Queensland. The project emerged from the aspiration of a local scientist and conservationist to build a research education centre that dealt with arid Australian environments and the preservation of the land, flora and fauna. The project concept had developed through community discussion, including with the local government. The centre was envisaged as having the potential for economic stimulation and employment opportunities, and a longer term impact on the township as a tourist destination within the outback. As the project took shape, the university and an architectural practitioner were invited to participate. Over the 2–3 years of the collaboration, the scheme led to the generation of ideas and a grant proposal attracting significant funding.

Pivotal Point

The valuable lesson arising was that unforeseen variables can lead to interventions that disrupt the process and aspirations as well as aborting the potential benefits for the community. In this instance it was a change of national government. The outcome was that the project was initially placed on hold, and then finally, funding was withdrawn under the new government, which meant a loss of over 5 million from the project and therefore the rural town. Economic and political imperatives came to dominate the outcome, regardless of the complexities and potential value of all other aspects.

Insights

- The advantage of being immersed within the town or community over time, rather than simply dealing with issues from afar, is multi-faceted. Immersion affords increased opportunities to observe and engage within the field and thus facilitates understanding through both informal and formal modes of engagement. Interaction and understanding are enhanced for all concerned.
- There will always be unforeseen variables that disrupt or abort aspects or the whole project. Resilience of both people and processes is an important aspect of a project, as well as planned contingencies being considered upfront and during its evolution.
- Economic and political imperatives can have high-order impact on any project. It is important to be aware and able to foresee and navigate these contingencies; selection of team members who can assist in building knowledge or can participate in negotiations, could be an advantage.

The Catalyst: A Checklist For Others

From these case studies, insights are critically analysed to provide reflections and ideas for a way of working in community engagement. These propositions acknowledge the limitations and potentials of any strategy for change in a complex environment, and may stimulate readers to reflect on new modes of working in their own future projects. Change can be as simple as restructuring the accepted process to include a more interactive and sensitive planning phase to explores the issues noted already within this chapter.

Nine broad categories have arisen for consideration: interaction, context, communication, attitude, scale, idea generation, external forces, forward-planning, impact. In addition to my own reflections, those colleagues interviewed raised issues arising from their experiences with other community engagement projects. Collectively, these considerations are captured in Table 12.3 as a list of suggestions or hints for practitioners (and others) who embark on community-based projects.

Broad observations include:

- Being aware, before embarking on a project, of the particulars of the specific context and the forces at play is very important;
- Having integrity when dealing with the project/research team and the community is of importance; and
- The project is an opportunity for exchange in many different forms. The experiences of community-based projects lead to friendships with locals, a chance to be in urban, rural or remote settings, and over time, a sense of familiarity and, in some ways, belonging, can develop, albeit not in the sense of being *a local*.

This last phenomenon (associated with trust and friendship) I have observed impacts in two ways in a range of projects. Firstly, the researcher/collaborator may try

Table 12.3 Suggestions for community engagement projects

Consideration	Facet	Practice [description, action and/or behaviour]
Interaction	Expectations	Friendships built with community require expectations and equality of all participants to be ensured
	Fractions	Recognize all communities are fractions-some operating with good will toward others; others reflecting tensions
	Outsider access fragility	Respecting access is Fragile and the real and potential consequences of being 'inside the fence' or Locked 'outside' needs to be considered at the outset
	Conflict: community members	Be mindful that during the process areas may be problematic such as people who arrive at workshops, buildings, etc. causing conflict
	Outsiders; Security + safety	Researchers are foreign regardless of good intentions and preparation. Being street wise or knowing the game will require time and knowledge and can involve both safety and a sense of personal security. Recognize researcher issues and emotions beforehand and manage; Researchers who are at ease engage more openly with the participants
	Engagement support	Involve people who were trained in working with the users with special needs so the latter can fully engage safely
	Empowered engagement	Engagement is facilitated when the research/community participants have ownership over the project from its beginning
	Passion	An important aspect of success is passionate staff, who are willing to go above and beyond; Good will and enthusiasm by all participants increases success
	Volunteering	Voluntary-based participation facilitates increase interest and engagement
	Collaborator status	Moving from the researcher/outsider to collaborator needs facilitation, and Logistical considerations may interfere with perceptions and acceptance
	Community; Immersive	Immersive engagement facilitates informal and formal modes of engagement, which enhance interaction understanding by all concerned
	Resilience	Resilience of both people and processes and planned contingencies is an important aspect as there will always be unforeseen disrupting variables
Local context	Fractions Interest group needs	Interest groups or fractions are not foes of a project. They are just that—people belonging to collectives, who prioritize one set of interests over others. The ability to recognize the Level of flexibility for a group and/or the means and sensitivity required to facilitate the acquisition of greater understanding between groups is also a valuable attribute to consider for a community engagement team

Table 12.3 (continued)

Consideration	Facet	Practice [description, action and/or behaviour]
Context priming	Site	Links to other existing projects in place can assist acceptance. If they have an established site, for example, it reduces overall cost of new project
	History: attitudes	An established history and legacy of the intention and the importance of the cause or the project means people and government have topic knowledge and can support immediately
	Honesty	Honesty of communication and agreed rules of interaction, including disclosure, may assist in more complex issues being addressed and outcomes achieved
	Communication: filters	Having a lead organization will lead to a filtered form of engagement. As a consequence, when building teams or partnerships attention to the building of open communication and exchange of ideas, as well as frank discussions regarding processes, may help to mitigate difficulties arising through filters applied by all of the players
Communication	Communication: sensitivity to diverse people	Having a team that can empathize and communicate with the diverse nature of the community can enrich the project
	Communication: authenticity	Authenticity impacts on the members of the entire project—community and researchers. When people genuinely wish to contribute and to facilitate change or improvement there is risk that a lack of authenticity can undermine meaningful relations and other outcomes that may have evolved if the full picture is understood

to tighten the process to make sure that something is seen to be being done. This may come from a sense of responsibility or obligation. In contrast, the ability to trust the process and to let things unfold by working openly with the community members may occur. This second alternative naturally involves more risk, as the outcome cannot be foreseen, even if the likely outcomes and impact are predicted.

The lessons learnt through experience are profound and have informed future attitudes, processes, and work practices. Also, awareness has grown of the need for contextually appropriate strategies; means by which one can complete a project and extract oneself from the context whilst honouring the commitment and expectation generated over the course of the project by all stakeholders.

Through these reflections, drawn from many years across very different projects, two general observations have arisen. Firstly, there are common threads running through very diverse projects, which may not be immediately evident. Secondly, through reflection-in-action and reflection-on-action we are able to learn from ourselves and from others, and these insights will inform future action. Community project activities and practices can be modelled on Donald Schön's concept of the reflective practitioner, as a means to improving the level and quality of engagement, and thereby, the potential impact of the projects.

References

Bertolini, L., Laws, D., Higgins, M., Scholz, R. W. Stauffacher, M., Van Aken, J. E., et al. (2010). Reflection-in-action, still engaging the professional? *Planning Theory & Practice, 11*(4), 597–619. doi:10.1080/14649357.2010.525370.

Billett, S. (Ed.). (2010). *Learning through practice: Models, traditions, orientations and approaches*. Dordrecht: Springer.

Bourdieu, P. (1986). The forms of capital. In J. Richardson (Ed.), *Handbook of theory and research for the sociology of education* (pp. 241–258). New York: Greenwood.

Gero, J. S., & Kannengiesser, U. (2008). An ontology of Donald Schön's reflection in designing. *International Journal of Design, Sciences & Technology, 15*(2), 77–90.

Goffman, E. (1959). *The presentation of self in everyday life*. Garden City: Doubleday.

Hagevik, R. Aydeniz, M., Glennon Rowell, C. (2012). Using action research in middle level teacher education to evaluate and deepen reflective practice. *Teaching and Teacher Education, 28*(5), 675–684. http://dx.doi.org.dbgw.lis.curtin.edu.au/10.1016/ j.tate.2012.02.006. Accessed 30 Aug 2013.

Moon, J. A. (1999). *Reflection in learning and professional development: Theory and practice*. London: Taylor & Francis. http://www.amazon.com/Reflection-Learning-Professional-Development-Practice/dp/074943452X/ref=sr_1_1?ie=UTF8&qid=1377411728&sr=8-1&keywords=Reflection+in+Learning+and+Professional+Development%3A+Theory+and+Practice. Accessed 30 Aug 2013.

Schmidt, M. R. (2000). You know more than you can say: In memory of Donald A. Schön (1930–1997). *Public Administration Review, 60*(3), 266–275.

Schön, D. (1983). *The reflective practitioner: How professionals think in action*. New York: Basic Books.

Schon, D. (2001). The Crisis of Professional Knowledge and the Pursuit of an Epistemology of Practice. *The HEC Archive*. http://www.heacademy.ac.uk/assets/documents/resources/heca/heca_cl13.pdf. Accessed 30 Aug 2013.

Chapter 13
Reflections on Working with Communities and Community-Based Projects in Bangladesh

Mokhlesur Rahman and Bob Pokrant

Abstract Community-based studies are a central feature of current development and climate change research and planning in developing countries. In this chapter we reflect upon our engagement with place-based communities over two decades in rural Bangladesh. We draw upon several encounters with local communities to illustrate the relationship between communities of practice and of place; complexities of putting what local people say and do at the centre of project work; the relationship of policy to practice; the difficulties of balancing immediate gains with longer term sustainability; and the need to locate place-based communities within wider ecological, economic and political networks of activity and influence to ensure more effective planning.

Keywords Community · Climate change · Adaptation · Bangladesh · Planning · Sustainable livelihoods

Introduction

The concept of community is an elusive one. It is used in a variety of ways in many disciplines, and debates continue about its meaning and usefulness as a substantive concept, a heuristic tool and as an appropriation device (Angerbrandt et al. 2011; Blackshaw 2009; Crow 2012). Substantively, Blaikie (2006, p. 1944) comments that 'community' has been used '… as a spatial unit, as a distinct social structure and as a set of shared norms' and that it denotes particular sets of social and cultural

B. Pokrant (✉) · M. Rahman
Department of Social Science and International Studies, Curtin University, Perth, Australia
e-mail: B.Pokrant@curtin.edu.au

M. Rahman
Center for Natural Resource Studies, Dhaka, Bangladesh
e-mail: mokhles_cnrs@yahoo.com

R. Tiwari et al. (eds.), *M² Models and Methodologies for Community Engagement*,
DOI 10.1007/978-981-4585-11-8_13, © Springer Science+Business Media Singapore 2014

relationships which may or may not be confined to spatially fixed and concentrated collectivities. The concept is tied to notions of identity and people's senses of place, which can and do transcend the common sense notion of place as physical location within some narrowly defined spatial and socially meaningful entity (See also Agrawal and Gibson 2001; Gupta and Ferguson 1997). It has been important in the broad field of development and globalization studies, particularly in the area of community-based natural resource management (CBNRM) with a focus on the mutual constitution of social and ecological systems (Berkes and Folke 1998), by which is meant that the social and the ecological can only be understood through their relationality. Our own interest in community derives from 20 years of research and development in geographically specific place-based communities in rural Bangladesh, on a range of issues such as wetland restoration, fisheries, aquaculture and natural resource management.[1]

During the years we have been working with place-based communities in Bangladesh and elsewhere, there have been shifts in the theoretical and policy focus of development studies and the place of community-based research within that focus. From the 1950s to the 1970s, development was seen as a replication of 'Western' paths to the higher living standards of the 'developed' world. The beneficiaries of development were seen as the tradition-bound peoples of the underdeveloped or developing world, who were to be freed from the dead hand of tradition and brought into the modern world of industry, democracy, and educational attainment. The perception was that these communities would benefit through imitation of generalized models of the historical development trajectories of rich countries. Such a process involved discarding values and ways of behaving legitimated by 'tradition' (the self-conscious reverence for what had gone before) and replaced by progress and development centred around values of achievement, merit-based aspiration, scientific expertise and a belief in the essential unboundedness of human potential (Scott 1998).

Much development thinking in western universities and institutions at that time was aimed at assisting developing countries to follow capitalist and liberal democratic notions of modernity. The key drivers of such progressive change were the political, economic and academic elites of the rich countries. Local communities were seen as the recipients and beneficiaries of their actions, rather than active agents of change.

---

[1] Rahman is a Bangladeshi environmentalist and community based natural resource management specialist who began his career in aquaculture while working for DANIDA, Caritas and FAO during 1983–1991, followed by work as a wetland ecologist with the Bangladesh Flood Action Plan in 1991–1994, and in community-based natural resources management (wetlands and forests) with various donor supported projects since 1994. His recent work has been on disaster management and climate change adaptation planning and management with USAID, DFID, UNDP, JICA, SDC, IUCN and Oxfam. He is currently pursuing PhD research into coastal communities' perceptions and adaptation to climate change impacts in Bangladesh at Curtin University, WA.

Pokrant is an Australian environmental anthropologist whose interest in natural resource management developed in the 1990s when he worked on the impact of British colonial rule on fisher communities in India, the organization and development of fisher communities in Bangladesh and West Bengal and the growth of an export-oriented shrimp industry in Bangladesh and its impact on rural communities.

In the 1970s, researchers and policy makers began to redefine local communities as potential reservoirs of entrepreneurship, best exemplified by the shift from talking about the traditional sector to the informal sector (Hart 1973; International Labour Office 1972; Kitching 1982). From being obstacles to change, local communities became the locus of change. Thus began the move towards a growing valorization of the idea of community and related concepts such as decentralization, participation, consultation, and local knowledge.

With the collapse of communism in 1990 and the growing dominance of neo-liberal ideology in the rich world, the term ‘community’ took on added significance as it reflected the new emphasis on reducing state direction and control of development and its replacement by increased private/corporate investment, export-led industrialization, free markets, a greater emphasis on governance as distinct from government, and an enhanced role for neo-liberal notions of civil society in the development process. Neo-liberalism stresses the re-configuring of ‘community’ to ‘free’ people from the constraints of government control and anti-competitive community practices, to allow them to pursue their own economic and other interests as essentially private individuals consisting of a bundle of property attributes. Partly in response to the neo-liberal focus on private property as a condition of community development, an academic and policy counter-narrative emerged in the form of community management of common pool resources,[2] in which local peoples engaged in forestry, fisheries and other natural resources were encouraged to adopt community property management principles to enhance productivity, equity and sustainability (Ostrom 1990, 2001; MacCay and Jentoft 1998).

The authors’ engagement with Bangladeshi rural communities developed during this period of neo-liberal ascendancy and was influenced by the parallel shift in approaches to natural resource management from state-centred, top-down models to those which sought to devolve decision-making to local peoples as beneficiaries of the projects (Gruber 2010; Stoll-Kleemann et al. 2010). Bangladesh has been a proving ground for community-based approaches to development and environment, as it is under increasing pressure from a range of natural and human-induced hazards, which include growing urban and rural populations, multiple land and water use demands, industrial expansion, infrastructure development, over-fishing, deforestation, terrestrial and aquatic pollution, cyclones, storm surges, flooding, coastal and river erosion and habitat loss, and climate change (Millennium Ecosystem Assessment 2005; United Nations Development Programme (UNDP) 2013; United Nations Environmental Programme (UNEP) 2012; Valiela 2006). It has also been heavily influenced in certain economic sectors (e.g., garments and fish/shrimp) by neo-liberal practices, which in the case of fisheries and aquaculture, have often clashed with attempts to institute common property regimes in various aquatic environments.

[2] ‘Common-pool resources (CPRs) are natural or human-made resources where one person’s use subtracts from another’s use and where it is often necessary, but difficult and costly, to exclude other users outside the group from using the resource’. Digitial Library of the Commons (Accessed April 20 2013. http://dlc.dlib.indiana.edu/dlc/contentguidelines

In this chapter, we reflect upon our personal and professional engagement over 20 years with a number of projects in selected rural communities in the areas of community-based natural resource management (CBNRM), community-based adaptation to climate change (CBA) and export-oriented shrimp farming. Each of us has worked with several place-specific rural communities that are dependent upon on the extraction and cultivation of local natural resources in the form of wetlands, fisheries, aquaculture, salt working and farming. The projects were funded by international donors (Ford Foundation and USAID in collaboration with the Bangladesh-based Center for Natural Resource Studies (CNRS)) and by external research agencies (a 3-year Australia Research Council grant in collaboration with the Bangladesh Centre for Advanced Studies (BCAS) and a 3-year Curtin University Strategic Scholarship in collaboration with the CNRS). The donor agency-funded projects focused on linking ecological conservation and sustainable livelihoods objectives, while two were more academic in orientation with potential policy implications, and focused on community adaptation to climate change and on the impact of the domestic shrimp value chain on local communities and environments. Figure 13.1 shows the project locations.

Communities of Practice and of Place

It is often assumed that it is only the people studied or helped who have community, while those who research, advise and assist them are disembodied experts who come from some cosmopolitan world, which allows them to distance themselves reflexively from their own cultural worlds and those they study (Harper 2011). However, project staff are themselves members of communities of practice (Wenger 2011) with their own 'professional habitus' (Mosse 2011, p. 22). Thus, we brought to our encounters with local people presuppositions, frames of understanding and sets of research skills derived partly from previous work with other communities, from our own disciplinary orientations and from our involvement with other communities of practice. In a sense, our engagement with local communities began before particular sites were selected and visited. In addition, we worked according to schedules set by the project parameters, funding arrangements and our own professional trajectories. For the most part, we did not treat the professional communities to which we belonged with the same critical eye as we did the communities studied. Put another way, we did not look at ourselves as 'ethnographic subjects' engaged in the process of making meaning and being engaged in imposing meaning on the place-based communities of interest.

Both authors were familiar with the project communities they studied through earlier visits of different length. During the actual projects, we lived and worked for variable and fragmented time periods from 1 week to 3 months at various field sites. Local community members regarded us as outsiders (Rahman from Dhaka and Pokrant from Australia) who came and went according to project schedules.

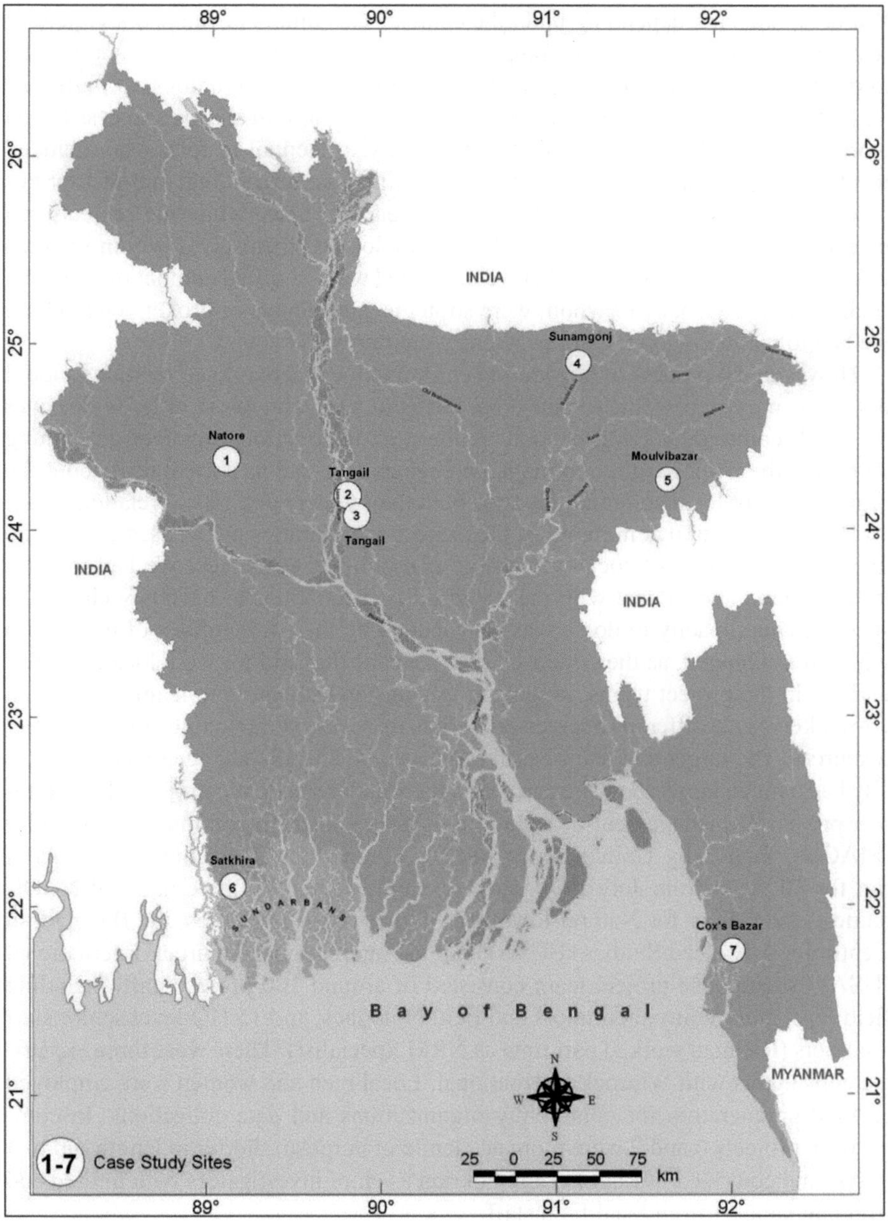

Fig. 13.1 Bangladesh map showing case study sites

While on-site, and depending upon the context, the authors developed a variety of relationships with local people ranging from regular and daily public and private interactions, predominantly with male members of the communities, to more distant interviewer-interviewee relations of limited duration. In all cases, access to and work within field sites was mediated by locally influential people, who included local Union Parishad members, businessmen, imams, district and sub-district bureaucrats, and occupational and community leaders. Thus, while the projects can be seen as community-based in so far as they located themselves within selected villages and regional towns, in fact we interacted with a limited number of people in each community, many of whom were strategic gatekeepers or brokers with whom we had to maintain good relations to carry out our work.

However, the process of negotiating entry into local networks of relations should not be seen as an obstacle to entry that needs to be overcome in order to discover the 'real' community lying beyond. Rather, the negotiations involved in gaining entry and the work required to maintain engagement told us a great deal about the organization of local communities and about their wider networks of relations. Like all Bangladeshi rural communities, the project communities were based around the unequal distribution of social, economic and political resources. We had to learn to understand and to deal with these and other inequalities to have any chance of success. Our capacity to do so was also shaped by the communities of practice to which we belonged, as they themselves were stratified and we were located differentially in the project teams, ranging from chief investigators to team leaders and co-workers with different degrees and types of access to and control over project resources. The larger and team-based projects with specific change objectives usually had a much larger local presence than the smaller, academic projects. For example, project 3 (management of aquatic ecosystems through community husbandry (MACH), funded by United States Aid for International Development (USAID) ran for 10 years from July 1999 and was implemented by three national NGOs, namely, the Center for Natural Resource Studies (CNRS) Caritas, the Bangladesh Centre for Advanced Studies (BCAS) and one international, Winrock International (USA) as lead. The project team consisted of around 100 project staff, including field level community facilitators and field biologists, and 15 HQ level senior staff members (Rahman worked part time as NRM Specialist). There were three expatriate consultants with Winrock International. Local men and women were employed as field enumerators for community organizations and data collections. In comparison, projects 6 and 7 were more academic in purpose, shorter in length (up to 3 years) and consisted of Rahman and Pokrant as chief investigators with limited and irregular support from local field staff.

We noted earlier that we brought to the projects some knowledge of how local communities worked, acquired prior to actual physical engagement with them through reading the academic and policy-related literature, briefings from colleagues and the like. Once on site, our capacity to understand the rhythms of everyday life and to participate in people's life-worlds was shaped by project size and length, specific professional tasks to be accomplished, language and culture, and status within the project and in relationship to community members.

Putting Local People First

At first glance it would seem obvious that, in a project which focused on place-based communities, local people who made up those communities would be central to the project task. However, putting local people first can mean many things, depending on project objectives, project methodologies and the contingencies of everyday life. Local people can take a strong deliberative role and be co-producers of the project design, implementation, evaluation and any follow-through. They can also be simply beneficiaries, 'objects of study' or sources of data. Over the past 30 years there has been a theoretical and policy shift from the latter towards the former (Cooke and Kothari 2001; Hickey and Mohan 2004; Mansuri and Rao 2012), although the extent to which this has happened in practice and whether it has resulted in better policy outcomes continue to be the subject of debate. In addition, projects often target particular categories or groupings of people, such as women or children, specific occupations such as fishers or farmers, or local organizations such as Non Government Organizations (NGOs) or local governments. The projects described here varied in these regards, with a trend towards increased participation and deliberative decision-making on the part of some members of the local communities over the 10–15 year period of the donor-funded projects. Increasing emphasis was placed on institutional and knowledge collaboration between the two communities of practice and place, which resulted in a shift from a sectoral and expert-centred approach to a more systems-based understanding (Armitage et al. 2011).

For example, the Ford Foundation and USAID-funded projects (projects 1, 2, 3, 4 and 5) had clear practical objectives aimed at achieving a more sustainable relationship between conservation biodiversity and people's livelihoods, through the spreading of ecological and other benefits within particular target communities. The earliest, Project 1, which was a USAID-funded Flood Action Plan[3] 16 (16) (environmental study) in Natore, North Bangladesh, and which ran from 1990 to 1991, was a study project and did not have as its brief the formation and engagement of community-based organizations. However, it became clear as the project progressed that as fisheries and agriculture were the key resources systems, any management of wetlands and thus Community-Based Natural Resource Management (CBNRM) in wetlands must involve fishers and farmers. Experience from this project led to a move towards greater knowledge and institutional collaboration and was incorporated into a later CNRS Ford Foundation assisted project in mid-1994.

In the CBA and shrimp studies, we were not official aid givers assisting local peoples to bring about changes in their local economic, political and social practices, such as establishing local committees, negotiating with 'stakeholders' to work out mutually satisfactory solutions, monitoring the effectiveness of new institutional

[3] The Flood Action Plan was a long term programme funded by the World Bank designed to protect floodplain and other populations from flooding. It consisted initially of 26 components and focused largely on 'hard engineering' solutions. For details of the Plan, see http://econ.worldbank.org/external/default/main?pagePK=64165259&theSitePK=477916&piPK=64165421&menuPK=64166093&entityID=000178830_98101904135218.

arrangements, and evaluating the practical outcomes for local communities. Rather, we presented ourselves as researchers interested in understanding and writing about local people's lives. As our project teams were small, we engaged more personally and intensively with a wide range of people within the local communities studied and further afield. In some respects, this made our engagement more difficult, as initially many did not see what practical value the projects had. We expended a great deal of time and energy in establishing our credentials through meeting local leaders, explaining our purpose, discussing project objectives, and convincing people to give up their time to assist the project.

Learning While Doing: from Policy to Practice

The donor-funded projects were change-oriented projects which were expected to deliver benefits to local people, and to do this while working with selected members of local communities.

Project 1: Listening to Local Voices

This was an 18 month project that focused on the impacts of water control structures on fisheries biodiversity and their effects on human nutrition, was part of a much larger USAID-funded project known as the Flood Action Plan-16 (FAP-16), which ran from 1990 to 1995. Rahman learned through his own experiences working with local people that changes were needed to how the project was organized to bring about any benefits. He worked in a team comprising national and expatriate consultants and field researchers (junior to mid level). FAP itself conducted various studies at the field level relevant to flood control and management, EIA case studies, wetland biodiversity, health and various socioeconomic studies. This study included an American Team Leader (part time), a Bangladeshi socio-economist, Rahman, who was hired as a fisheries/wetland ecologist/biologist, and 16 field researchers (8 female and 8 male).

The field researchers were field-based and had direct regular interactions with the local communities in their respective study sites. Their main tasks, performed mainly by female researchers assisted by male researchers, were to collect data from selected households on household daily consumption of rice, fish, vegetables, pulses, meat, eggs and other protein food items, including selected socioeconomic parameters. The male field researchers engaged in: fish catch surveys at selected fishing locations twice monthly for 12 months; fish migration studies in selected sites during early monsoon; selected fish market surveys for 12 months; Focused Group Discussions (FGD) and Key Informant Interviews (KII); and school children surveys on fishing and fish consumption, local wetland mapping and so on.

The HQ level consultants paid regular field visits to four sites to monitor, guide and quality check the work done. They also met with some household members to discuss fish consumption and carried out FGDs and KII surveys with household members and fishers at fishing locations to widen their understanding of the study issues and objectives.

In addition, the project had part time HQ-based data entry operators, a systems analyst with partial inputs from a statistician, a wildlife biologist, agronomist, engineer, and an expatriate economist from the FAP-16 team.

Rahman's experience taught him the central importance of working with local communities through a mosque-based committee (Community Based Organisation or CBO), which leased three state-owned ponds (*khas*) (called *pagars*) in a large low lying floodplain stocked naturally with diverse fish species during monsoon inundation. In the past, the floodplain was transformed during the monsoon into a thriving fishery where local people had relatively unregulated access to the ponds for fishing. With the recession of floodwaters in the late to post-monsoon period, the basin and its fishing stocks returned to private leaseholding and regulation by leaseholders of larger depressions, called *beels*, and by owners of numerous *pagars* dug on private lands to catch fish. The *pagars* were either fished out by the owners themselves or rented to fishers, resulting in the complete drying out of the *pagars* and removal of all fish.

The CBO used money earned from fishing for village development purposes, including helping very poor families at crisis times. It added a condition to fishing rights that fishers were not to completely dry out the *pagars* but retain waist deep water so that fish and other aquatic biota could survive over the dry season and allow natural repopulation of the floodplain for the next season. The CBO also sought to provide drinking water for cattle as well as for bathing and drinking for farmers who worked the land during the dry and hot summer. As a result, after the recession of floodwater, the entire flooded basin became a vast crop land during the dry season.

The actions of the CBO showed that communities can act to provide general benefits rather than simply serving the interests of one section of it. It was members of the local community rather than outside researchers who came up with the idea of community-based conservation, which demonstrated that local people were aware of social-ecological problems and were able to provide their own solutions without much support from external sources. The project itself learned the importance of listening to local voices and the potentiality of applying what had been learned to other community-based sanctuaries in other parts of the country.

Project 2: Applying Lessons Learned Within and Beyond the Local Community

This USAID-funded Flood Action Plan project-16 (Environmental Study) illustrates how the lessons learned from one project's experience were extended to another. It also illustrates the importance of seeing 'community' as embedded within complex ecological networks that sustain community livelihoods and extend beyond the presumed spatial boundaries of a single community.

In this project on traditional fishers in Tangail District, fishers' livelihoods depended on fish migrations from river to floodplain during early monsoon. The fishers lived on the bank of a state owned perennial *beel* (wetland) and fished the *beel* on adjacent seasonally flooded lands by the river. The *beel* was connected to the river by a narrow canal which acted as a migratory route in the early monsoon, allowing large quantities of fish eggs, tiny fry, fingerlings and adults to reach the *beel* and seasonally flooded lands for breeding, feeding and growth.

Initially, the project focused on selecting a point on the canal where set bag fishing nets could be placed to trap migrating fish so the research team could sample and code data on immigrating fish flows. However, in that year, early monsoon rains were inadequate, and the canal dried up and became silted as a result of the deposition of sands on the canal bed and at its mouth with the river. Rahman observed that the fishers of the village complained of the unavailability of mature fish and fish spawn in the *beel*, which they expressed with the phrase *pani nai, mach nai* (no water, no fish). This observation led Rahman and his colleagues to recommend the rehabilitation of silted canals to allow free movement of fish.

Project 3: Local Change Agents and Community-based Organizations

The community and researcher observations made during Project 2 were incorporated into Project 3, a Ford Foundation-funded, community-based wetland habitat restoration project in Tangail District from 1994 to 1996. This project illustrated the problem of free riding in a common-pool resources context.[4] The project aimed to rehabilitate a canal that linked a local floodplain with the nearby river and monitor the changes. An *ad hoc* local project implementation committee (PIC) was formed, the canal was desilted using local labour, and fish movements monitored and documented with the results provided to local people, government, World Bank, the UK Department for International Development (DFID), Worldfish and the Ford Foundation. Local household consumption of fish was also recorded. In order to measure changes in fish production and composition from canal rehabilitation, the project trained and engaged local boys and girls with some literacy and numeracy skills to assist in the impact monitoring program. These enumerators improved their knowledge base and turned into local change agents who acted as advocates of fish conservation among other members of the community. These young people sensitized project staff and local people to the need for other kinds of conservation measures to sustain the restored fisheries. These were: a ban on the use of harmful fine mesh seine[5] nets during fish breeding; no fishing in the canal during specified

[4] Free riding refers to: '…the act of freely using resources generated by another person without payment or even authorization…' (Frischman 2013, p. 13).

[5] Seine nets are used on beaches and on boats. They are long nets with or without bags that encircle an area where fish are found. They have been used for centuries among inland fishers in Bangladesh.

periods to facilitate immigration of fish from rivers to wetlands; and establishing small sanctuaries. The communities were not ready to accept these limits initially.

The research assistants later reported that people were catching migrating fish at the canal/river intersection, after which a tripartite meeting of set bag net fishers, the PIC and the research team agreed that during the pre-monsoon period, which coincided with peak fish migration, no set bag net fishing would be allowed, but other fishing gear could be used under specific circumstances.

However, the project also learned that several groups of fishers from adjacent villages had started fishing with fine mesh seine nets in the recently inundated floodplains, catching tiny fish fingerlings and gravid adults of diverse species spawning in productive flooded lands. This type of fishing was not included in the original project design, as it was not recorded in the baseline survey.

A wider cross-community level meeting was held, and it was agreed to ban use of seine nets for the first 2 months from the date of inundation of floodplains. Fishing with all types of gear was permitted as long as there was sufficient rainfall for 4–5 months from June to October. The ban on seine nets allowed the growth of submerged aquatic vegetation in the floodplain, which obstructed seine net fishing in at least 40 % of the flooded basin and provided a safe haven for fish. Catch data showed a large increase in catch per unit effort and species diversity grew. In addition, the PIC and project team met with private *pagar* owners to select one *pagar* to conserve fish stocks during the post-monsoon period. As a nursing ground and safe haven for fish, it provided the whole water body with fish, benefitting all who fished during the monsoon period. At the end of the project year, fish catch data showed a six-fold increase in fish production from the baseline and an increase in fish species from 46 to 64 in year 2 (Rahman et al. 1999).

Several lessons were learned from this project. First, sustainable fish conservation required multi-pronged conservation intervention (namely, specific gear banning, time closure on fishing, wetland sanctuary) as opposed to the initial plan of a single activity of canal rehabilitation. Second, rather than having *ad hoc* community institutions (PIC) to carry out single and time-bound tasks (canal rehabilitation), permanent community institutions were necessary to ensure conservation continues beyond the life of a single project. Such institutions required engagement of a range of local resource users, including rice farmers who were the main land owners and exerted considerable power in rural society. Fishers themselves were not a homogeneous category and, as the example shows, had different and often competing interests and degrees of influence, which must be recognized to ensure effective collective decision-making. From these early beginnings, community-based organizations (CBOs) have now become an established part of natural resource management in Bangladesh (Thompson 2013).

Third, the project provided local capacity building and training for young people who were familiar with local conditions and who could work with local people in the longer term. Fourth, a potential clash of competing property rights notions was reduced through negotiation with holders of water rights, which allowed a more equal access to aquatic resources. However, the project also showed the difficulties of ensuring local control over a resource when the ecological systems underpinning resource use extend across several communities. The systems of property rights themselves reflected wider political structures of power and influence extending to the national level, which provide political constraints on the actions that single projects could take. Fifth, any project must be open to contingency, unseen events and processes that cannot be predicted beforehand. Local people can deny, dissemble, alter and transform project objectives, as well as assist project workers to realize project objectives (Mosse 2005). Finally, as a result of this experience, new ideas and practices were generated that could be used in other projects. Thus, Rahman's experience and growing expertise allowed him and some colleagues to establish an environmental NGO named the Center for Natural Resources Studies (CNRS), which tested and implemented various lessons learned from the FAP study on fisheries. CNRS developed a concept of 'community-based wetland habitat restoration and fisheries management' which was funded by the Ford Foundation in two phases for 7 years starting from July 1994, and demonstrated the concept in four locations. Later UNDP and USAID adopted the idea that CNRS demonstrated on a small scale, and through the Department of Fisheries (DOF) and the Department of Environment (DOE) up-scaled the concept to other floodplains of Bangladesh. A CNRS demonstration project on community-based wetland habitat restoration and fisheries management was later up-scaled in Bangladesh in a number of projects funded by different agencies and implemented in different floodplains. Since this time, CNRS has been actively involved in various donor-supported government projects on fisheries and wetland management.

Project 4: Community-based Organizations: Restoring the Balance of Ecosystems

The Ford Foundation project was extended to fishing and other communities in the Haor region of Northeastern Bangladesh, a vast, low-lying, depressed basin bordering the Indian Meghalyan hills. The region is under water during the monsoon and is a major fisheries resource, while during the dry season, the entire basin is turned into a vast crop land which, with the exception of a number of perennial wetlands, is used for farming winter rice. The wetlands are sources of fish, which attract large numbers of migratory birds and water fowl, making it an internationally important biodiversity area, parts of which have Ramsar wetland status.[6] Initial meetings with local people showed that their main concern was the declining balance between forest and wetland, expressed with the comment: *jangal nai, mach nai*, or, no forest, no fish. They explained that in the past there were large areas of raised lands (*kanda*) along the edges of wetlands full of swamp forests (trees, reed lands and other shrubs and herbs). During the monsoon flooding, diverse species of fish colonized the

[6] A Ramsar wetland is an ecosystem protected under the International Convention on Wetlands of International Importance, more commonly known as the Ramsar Convention. See http://www.ramsar.org/cda/en/ramsar-home/main/ramsar/1_4000_0.

flooded forests to breed, nurse and feed, and small prey fish species took shelter to escape predators. Those areas with greater swamp forest coverage had higher fish productivity. Over recent decades, people had settled in the region and cleared the forest for farming, which reduced forested *kanda* areas. The project's aim was to assist in redressing the imbalance by conserving the remaining *kanda* areas. However, conservation involved fishing restrictions at certain times, which led local fishers to seek micro-credit support to help find alternative work during such times. This was not part of the original project brief, but this changed after negotiations with the donors. The project planted denuded *kanda* lands with two swamp tree species in three sites, and trained communities to be aware of the value of fishing time closures, specific gear bans, habitat restoration and wetland sanctuaries to enhance fisheries production and the value of wetland biodiversity. This was achieved through village level community based organizations (CBOs) acting as sustained community spaces to facilitate local level conservation.

Project 5: Elite Capture and the Political Ecology of Wetland Management

Another pre-designed USAID-funded MACH project (Management of Aquatic Ecosystems through Community Husbandry) began in 1999, and incorporated lessons from past and ongoing projects. Rahman was one of the two project designers; he played a key role in the implementation of the project, and his engagement was affected by community suggestions. Initially, project interventions were largely confined within the spatial boundary of wetlands in the Hail Haor basin of Mulvibazar District. The Hail Haor wetland covers 12,500 ha of land surrounded on three sides by small hills, and receives water annually from the hills through narrow streams, which peak in the monsoon. During the first year, community advice provided during action planning sessions was to widen the spatial and social boundaries of the project to include the watershed and the hills. The rationale was that the hill slopes were used for what local people considered unsustainable pineapple and lemon cultivation, which diverted water from the hill streams and facilitated massive soil erosion which, in turn, raised the wetland beds in the Haor. Moreover, stream bank stabilization was necessary to arrest soil erosion further down the hill slopes caused by forest clearing along the banks. In the second year, through participatory assessment of watershed-related issues and discussions with the hill slope cultivators, the project was able to widen the spatial boundary and include the watershed, and launched watershed management interventions. These were: contour pineapple cultivation; stream bank protection through large-scale riparian plantations; crop diversification to reduce water requirements for crop irrigation; dialogue with tea estate managers to plant trees along the streams passing through their estates; and the formation of watershed management committees as local institutions linked to the agriculture extension department.

While community input was valuable in bringing about positive conservation outcomes, a problem arose of what has been called 'elite capture' (Platteau 2004; Wong 2010). Haor communities are highly stratified along class and ethnic grounds, and local elite groups who had benefitted earlier from legal and illegal fish leasing arrangements persuaded some villagers that the USAID project would exclude them from the benefits of such leasing arrangements. They were told that the Americans were aiming to take over the wetlands and prevent local people fishing. This delayed initial planning activities, but local agents (members of the project who organized poor fishers' groups) arranged a series of courtyard level meetings, at which the targeted communities were persuaded of the positive benefits of the project. This highlights the central importance of having a detailed understanding and on-going monitoring of the structure of power relations in local communities, which can extend to district and national level politics. One way to reduce elite capture is to ensure all village-level committees have a majority of the less well-off members of the community. However, this does not guarantee more distributive or procedural justice payoffs, as elites can act informally and committee members may make decisions with elite reactions in mind.

After 8 years, the project was able to establish the only large wetland sanctuary in the country—Baikka Beel sanctuary—which created a safe haven for fish, water fowl and wildlife, and a site for wetland based ecotourism. However, more recent events illustrate how fragile such sanctuaries can be. Thus, in April 2013, during the pre-monsoon season, the area was affected by prolonged drought and high temperatures, which lowered water levels in the sanctuary, led to decomposition of aquatic vegetation and increased oxygen depletion, resulting in high fish mortality. People from villages close by sought to fish out the sanctuary. This illustrates the sensitivity of CBOs to changing environmental conditions, and the importance of understanding the political ecology of natural resource-dependent communities and how economic, social and political networks extend across ecological and spatial boundaries.

The Interpenetration of the Local and the Global: Adaptation to Climate Change and Shrimp Exporting

More recently, the authors have engaged in two academically-oriented projects on adaptation to climate change (project 6) and on shrimp farming in coastal Bangladesh (project 7).

Project 6: Engagement in a Multi-sited Project

This project on the globalization of the Bangladesh's shrimp-export sector illustrates the challenges faced by an overseas researcher working across several place-based communities and organizations that are linked by economic and other networks, extending from remote rural locations to industrial zones of large cities. Over 3 years, Pokrant spent several weeks and months at a time in Bangladesh and in Europe interviewing, observing and socializing with: shrimp farmers; shrimp hatchery owners and workers; shrimp fry collectors; shrimp and fry traders; depot owners; salt workers; processing plant owners, managers and workers; processing plant labour contractors; government officials; and NGO workers. He collaborated with university and NGO colleagues, and employed several research assistants (two of whom were from the local shrimp farming areas studied) who continued to carry out project tasks when Pokrant was away from the research sites.

This project was a departure from Pokrant's previous research, which had focused in and on place-based communities. By shifting to a multi-sited ethnographic approach (Coleman and Von Hellermann 2011; Sorge and Roddick 2012), he was able to examine historically and contemporaneously how farming communities that had previously been relatively spatially isolated had become integrated into global food chains, and the impacts of this integration on the social and economic lives of local people and the ecological environments in which their communities were embedded.

A major challenge was the fragmented and multi-sited nature of the field work, which affected the researcher's capacity to observe the rhythms of everyday life of particular communities over an extended period, including the shifts from shrimp farming to salt production as the seasons changed. Also, his limited Bengali language skills, particularly in the early stages of the project, meant he relied upon assistants to record conversations when he was present or absent. Project funding did not allow investment in a large team of highly skilled local researchers, although Pokrant did work closely with a local sociologist who had knowledge of shrimp farming. A great deal of time was spent translating interviews and informal conversations into English, checking for gaps in information, and re-interviewing.

Pokrant also interviewed processing plant owners, overseas buyers and European Union food safety officials; this was necessary to piece together the ways in which the domestic supply chains were linked to the international market for shrimp. Corporate confidentiality concerns meant that managers and owners were sometimes reluctant or unwilling to provide sensitive information on product prices, food safety standards, and contractual arrangements. Such 'studying up' altered the dynamics of engagement, as Pokrant was in several instances granted limited time to speak with senior and middle managers in large companies, and 'come-back' interviews were difficult or impossible to arrange. However, in smaller shrimp distributive companies, more informal arrangements were possible and Pokrant was able to socialize, albeit in a limited fashion, with owners. Overall, this part of the research generated the least amount of useful data compared with the time and money invested, and would have benefitted from a co-investigator able to take on some of the interviewing tasks and to provide other perspectives. In this way the field work would have been both multi-sighted and multi-sited.

The global trade in shrimp was an internationally contested trade, and Pokrant spoke with anti-industrial shrimp environmental groups in London and in Bangladesh, who were ready to provide a great deal of information on their objectives and organizational structures. The global and contested nature of the shrimp export sector had created much suspicion among processing plant owners and senior managers in Bangladesh about the motives of 'outsiders' asking questions about the industry. For an anthropologist used to studying people of lower socio-economic status, Pokrant found himself seeking interviews and access with politically well-connected plant owners who, in some instances, granted him an audience rather than agreeing to an interview. Some plant managers were suspicious of his intentions and either refused to be interviewed or agreed to meet for short periods only. However, some were cooperative and others became more so over time. Through persistence, frequent visits to plants, and support from the relevant government authorities, he was able to build up trust and move relatively freely around factory premises. He and his team were given permission to interview middle managerial staff and male and female shrimp processing plant workers in several plants in Chittagong. In this way he was able to research commercial links between the plants and shrimp suppliers and between plant workers and their rural origins.

Pokrant followed what Sorge and Roddick (2012, p. 278) have referred to as the traditional meaning of multi-sited research and 'followed the commodity' from brood shrimp capture in the Bay of Bengal to processing plant in the main urban centres of the country, and to shrimp importing firms, processors and European Union food safety and hygiene agencies in parts of Europe. In this way, he was actually uncovering complex social, economic and political networks between people differently and unequally located in the global political economy of seafood. At the village level he observed how local people's everyday life-worlds were being changed. For example, many rice farmers shifted to shrimp farming, which offered them higher incomes but undermined traditional sharecropping arrangements with landless labourers and marginal farmers, and forced shifts into other work and migration to other areas. Farmers became dependent on extensive international commodity chains over which they had no control and so were subject to the vagaries of international pricing and the control of the domestic and international networks by traders, processors, importers and foreign consumers. The expansion of shrimp farms led to increasing levels of soil and water salinity, which affected the capacity of rice farmers to continue farming. Land, which had been common property accessed by various social groups, particularly the landless, was appropriated by wealthier and more powerful locals and outsiders. There was a decline in land and water areas devoted to cattle grazing, horticulture and fishing. Some mangrove areas were devastated by shrimp farming, as the government leased out certain hitherto protected forests to shrimp investors. The expansion of the sector led to the growth of jobs outside the local communities, which resulted in out-migration to the cities and in-migration as people from poorer regions sought work in shrimp farming, shrimp fry collecting, transport, trading and the like. Local protests against land appropriation and displacement of local people took place, sometimes supported by national and local advocacy groups.

Pokrant's experiences on this project taught him the value of pre-field work preparation. Taking on such a large project as a single investigator meant he stretched his resources, time and energy across too many research locations and it would have been wiser to have had a co-investigator. This was especially important in interviewing government officials and company owners and managers in Bangladesh and Europe who were themselves members of other communities of practice with their own professional codes of right conduct, their own 'paraethnographic expertise' (Sorge and Roddick 2012, p. 292) and their own ways of dealing with non-members who sought to penetrate their professional worlds. Logistically, it was necessary to determine with some precision when such people would be available, as travel was costly and time-consuming. Some information was made available via email contact but Pokrant could not rely on regular responses from middle managers and company owners and senior managers were unwilling to provide details of company operations in this way. There was no substitute for being there, although there were many 'theres' in which to be and not enough time to be in them.

He also learned that managers, owners and officials belong to communities of place with their own cultural practices and life-worlds, which need to be studied in some depth to understand how such participants view the world and their place(s) within it. Interviews carried out in work places with busy people over short periods of time are insufficient to capture such complexities. These encounters, brief though they were, taught him the value of persistence and patience when relationships with those being researched became awkward and troubled (see Gallo in Coleman and von Hellermann 2012). The researcher has to get used to being ignored, re-directed, talked down to, preached to and in other ways marginalized.

Project 7: Strange Weather we are Having: Adapting to Climate Change

The final project discussed in this chapter is an on-going project that focuses on adaptation to climate change among coastal communities in Southwest Bangladesh. It falls within the broader framework of community-based adaptation to climate change (CBA) (Ensor and Berger 2009; Reid 2009) at the core of which is the identification of communities most vulnerable to climate change, with a view to assisting them to adapt to what are regarded as the inevitable impacts of climate change for such communities. The project itself is largely academic, but informed by the CBA discourse, and has potential policy implications for adaptation planning. It focuses on the inhabitants of two village communities located within the same agro-ecological zone, and how they understand and perceive weather and the wider issue of climate change. The two villages were chosen because, while located within 15 km of each other, they have been differentially impacted by various weather events and processes (the 2007 and 2009 cyclones being the most recent), and they have had different experiences and provided different assessments of the relevance of those impacts to their lives and livelihoods. These differences reflected differences in inter- and intra-village sources of vulnerability, as well as differential exposure to weather events such as cyclones and tidal surges.

A key project challenge working with the two communities was how to link slow acting changes such as climate change with their own priorities, which were more immediate and focused on sustaining their everyday livelihoods (Giddens 2009; Jasanoff 2010). We considered that a first step towards this goal was to understand: what local people with different gender, occupational and socio-economic backgrounds saw as the key problems facing them; what they considered to be possible solutions to such problems; and if any of these problems could be attributed to weather or weather-related events and processes. The authors decided to focus initially on what Jasanoff (2010, p. 238) refers to as '...the mundane rhythms of lived lives and the specificities of human experience...', not simply because this is a central ontological fact of human life, but also because, in the eyes of the people we worked with, it made it easier to get them to talk freely.

The research found that there is a high degree of awareness of changing weather patterns, such as irregular monsoon rains, higher sea levels, changes in seasons, more intense cyclones, and localized rough weather events in the Bay of Bengal. These changes impact differently on different occupational groups and across gender lines. For example, for rice farmers in one village, insect infestation is given high priority, while in the second village, fishers and honey collectors consider water pirates and corrupt forestry officials as their top priorities. While the language of climate change has begun to enter local vocabularies, for most villagers it remains an abstraction that is often used as a synonym for unusual weather. NGO workers and some local government officials, who might be expected to play a role as disseminators of knowledge about climate change, demonstrated that their knowledge was often superficial and incorrect. For example, some stated that oxygen is a greenhouse gas.

The research also revealed that local people are aware of some of the economic and political causes of their own social vulnerabilities, which range from the actions of local elites who take land and divert water for shrimp farming, to the actions of corrupt local officials. Villagers described how they have struggled over many decades with natural hazards such as cyclones, tidal surges, flooding and drought, and the many coping strategies they have adopted to deal with them and increase their resilience. These include short-term migration, alternative employment, embankment building, informal loan support from local money lenders, pooling of resources with kin, and extended social networks. They are also increasingly familiar with how their own livelihoods depend on actions taken not only by their government and international donor agencies, but by other states, communities and agencies.

Conclusion

We have learned a great deal over the past 20 years of working at the grass-roots level with diverse place-based communities in Bangladesh. One important lesson is to avoid blueprint solutions (Berkes 2007) and understand such communities within their specific historical, political, economic and ecological contexts. As the examples illustrate, this does not deny the possibility of knowledge transfer from one situation to another, but it cautions against 'one-size-fits-all' thinking. Doing research is an iterative process and lessons learned in one project were adapted to new projects as our knowledge accumulated.

Place-based communities are best seen as arenas of conflict, cooperation and diversity, and as components of wider social, economic and political networks that both limit and provide opportunities for action that extend beyond the local. Local ecologies and the wider political economy of resource use are mutually constituted, and understanding their interconnectedness is essential (Blaikie 2006). Researchers and project actors are themselves outsiders to the communities studied, and this shapes how we engage with those communities and how communities see us. Project implementation is primarily a political process in which different collectivities compete for control over and the benefits from the policy initiative (Mosse 2005; Gardner 2012). Project leaders and their support teams are themselves entangled in local power relations and struggles rather than being mere technical service providers. Local priorities often differ from project or state priorities, and researchers must be ready to re-think project directions in the light of experience. Giving local people a stake in project work is essential to project implementation, and researchers must be ready to see project objectives change, both as a result of increasing knowledge about the local circumstances, and also through the influence of local people wanting changes to how and for whom the project is being carried out. Project dependence can be avoided by building in institutional measures to assist local people to take control of project activities for the longer term, and for this to happen, a longitudinal approach and research design will work more effectively than projects with limited time-frames. However, the life-worlds of members of place-based communities, company owners, government officials and others are subject to the vagaries of everyday life in which the long-term often appears as abstract and less important, a problem likely to be exacerbated in the light of climate change.

References

Agrawal, A., & Gibson, C. C. (Eds.). (2001). *Communities and the environment: Ethnicity, gender, and the state in community-based conservation*. New Brunswick: Rutgers University Press.

Angerbrandt, H., Lindstrom, L., & de la Torre-Castro, M. (2011). What is this thing called 'community' good for? In R. Chuenpagdee (Ed.), *World small-scale fisheries: Contemporary visions*. Delft: Eburon.

Armitage, D., Berkes, F., Dale, A., Kocho-Schellenberg, E., & Patton, E. (2011). Co-management and the co-production of knowledge: Learning to adapt in Canada's Arctic. *Global Environmental Change, 21*(3), 995–1004.

Berkes, F. (2007). Community-based conservation in a globalized world. *Proceedings of the National Academy of Sciences, 1040*(39), 15188–15193.

Berkes, F., & Folke, C. (Eds.). (1998). *Linking social and ecological systems: Management practices and social mechanisms for building resilience*. Cambridge: Cambridge University Press.

Blackshaw, T. (2009). *Key concepts in community studies*. Thousand Oaks: Sage Publications.

Blaikie, P. M. (2006). Is small really beautiful? Community-based natural resources management in Malawi and Botswana. *World Development, 34,* 1942–1957.

Coleman, S., & Von Hellermann, P. (Eds.). (2012). *Multi-sited ethnography: Problems and possibilities in the translocation of research methods*. New York: Routledge.

Cooke, B., & Kothari, U. (Eds.). (2001). *Participation: The new tyranny*? New York: Zed Books.

Crow, G. (2012). Community re-studies: Lessons and prospects. *The Sociological Review, 60,* 405–420. doi:10.1111/j.1467-954X.2012.02091.x.

Ensor, J., & Berger, R. (2009). *Understanding climate change adaptation: Lessons from community-based approaches*. Rugby: Practical Action Publishing.

Gardner, K. (2012). *Discordant development: Global capitalism and the struggle for connection in Bangladesh*. London: Pluto Press.

Giddens, A. (2009). *The politics of climate change*. Cambridge: Polity Press.

Gruber, J. S. (2010). Key principles of community-based natural resource management: A synthesis and interpretation of identified effective approaches for managing the commons. *Environmental Management, 45*(1), 52–66.

Gupta, A., & Ferguson, J. (1997). *Culture, power, place: Explorations in critical anthropology*. Durham: Duke University Press.

Harper, I. (2011). WORLD HEALTH AND NEPAL. In D. Mosse, *Adventures in Aidland: The Anthropology of Professionals in International Development*.

Hart, K. (1973). Informal income opportunities and urban employment in Ghana. *The Journal of Modern African Studies, 11,* 61–89. doi:10.1017/S0022278X00008089.

Hickey, S., & Mohan, G. (Eds.). (2004). *Participation--from tyranny to transformation?: Exploring new approaches to participation in development*. London: Zed books.

International Labour Organization. (1972). *Employment, incomes and equality: A strategy for increasing productive employment in Kenya*. Geneva: International Labour Office.

Jasanoff, S. (2010). A new climate for society. *Theory Culture Society, 27,* 233–253. doi:10.1177/0263276409361497.

Kitching, G. (1982). *Development and underdevelopment in historical perspective: Populism, nationalism and industrialization*. London: Methuen.

McCay, B., & Jentoft, S. (1998). Market or community failure? Critical perspectives on common property research. *Human Organization, 57*(1), 21–29.

Mansuri, G., & Rao, V. (2012). *Localizing development: Does participation work?* Washington, D.C: World Bank Publications.

Millennium Ecosystem Assessment. (2005). *Ecosystems and human well-being: General synthesis*. Washington: Island Press.

Mosse, D. (2005). *Cultivating development: An ethnography of aid policy and practice*. London and Ann Arbor: Pluto Press.

Mosse, D. (2011). Introduction: The anthropology of expertise and professionals in international development. In D. Mosse (Ed.), *Adventures in aidland: The anthropology of professionals in international development*. Oxford: Berghahn.

Ostrom, E. (1990). *Governing the commons: The evolution of institutions for collective action*. Cambridge: Cambridge University Press.

Ostrom, E. (2001). Reformulating the commons. In J. Burger, et al. (Eds.), *Protecting the commons: A framework for resource management in the Americas* (pp. 17–41). Washington, DC: Island Press.

Platteau, J. P. (2004). Monitoring elite capture in community-driven development. *Development and Change, 35*(2), 223–46.

Rahman, M., Capistrano, D. A., Minkin, S. F., Islam, A., & Halder, S. (1999). Experience of community managed wetland habitat restoration. In H. A. Middendrop, P. M. Thompson, & R. S. Pomroy (Eds.), *Sustainable inland fisheries management in Bangladesh* (pp. 111–121). Philippinnes: ICLARM Conference Proceedings.

Reid, H. (2009). *Community-based adaptation to climate change*. London: IIED.

Scott, J. (1998). *Weapons of the weak*. New Haven: Yale University Press.

Sorge, A., & Roddick, A. P. (2012). Mobile Humanity: The Delocalization of Anthropological Research. *Reviews in Anthropology, 41*(4), 273-301.
Stoll-Kleemann, S., De la Vega-Leinert, A. C., & Schultz, L. (2010). The role of community participation in the effectiveness of UNESCO biosphere reserve management: Evidence and reflections from two parallel global surveys. *Environmental Conservation*, *37*(03), 227–238.
Thompson, P. M. (2013). Sustainability of community-based organizations in Bangladesh. *Society & Natural Resources*, *26*(7), 778–794.
Wenger, E. (2011). Communities of practice: A brief introduction. scholarsbank.uoregon.edu. Accessed 9 Jan 2013.
Wong, S. (2010). Elite capture or capture elites? Lessons from the 'counter-elite' and 'co-opt-elite' approaches in Bangladesh and Ghana, Working paper. World Institute for Development Economics Research, No. 2010, 82.
United Nations Development Programme. (2013). *2013 Human development report*. New York: United Nations Development Programme.
United Nations Environmental Programme. (2012). *Global environmental outlook*. New York: United Nations Environment Programme.
Valiela, I. (2006). Global coastal change. Malden: Blackwell Publishing.

Chapter 14
Action Preparedness Tool for Community-Based Projects

Clancy Read

Abstract Community participation in an action-oriented project is intuitively appealing. However, the numerous contexual challenges that arise during implementation can turn community-led action into an idealistic pursuit rather than a practical reality.

A shift in focus to one that identifies these potentially hindering contextual factors in a project setting could influence the likelihood of successfully transforming knowledge to action. The ability to predict whether efforts are likely to be hindered by contextual factors and, as a result, impede positive change in communities, can help to better direct limited resources; and importantly, can provide insights into why so many development efforts fail to achieve lasting change in communities.

This chapter presents a new tool, the 'Action Preparedness Tool for Community-based Projects' that has been formulated as an instrument to identify and categorize context-specific challenges that may arise during efforts to implement participatory action with communities. This tool, to be applied before investing time, resources and effort into a project, has application across a broad range of disciplines. It is intended to help stakeholders identify potential road blocks in their project, which will give them options to develop proactive strategies to address these barriers, or to realise that action on the specific issue may not be feasible.

Keywords Community participation · Contextual challenges · Action · Participatory research

Introduction

This book showcases a variety of community-based projects from a range of disciplines. A commonality between the chapters is that whilst community participation is a worthwhile aspiration, it requires enormous commitment. Despite best in-

C. Read (✉)
Centre for International Health, Curtin University, Perth, Australia
e-mail: clancyread@hotmail.com

R. Tiwari et al. (eds.), *M² Models and Methodologies for Community Engagement*,
DOI 10.1007/978-981-4585-11-8_14, © Springer Science+Business Media Singapore 2014

tended efforts, many barriers to achieving true participation in an action-oriented project exist; and in the case of Read (2012), can hinder and even halt efforts to transfer knowledge of community problems into action to create lasting change for communities.

This chapter provides a continuation of learning from Chap. 6, Applying a Practical, Participatory Action Research Framework for Producing Knowledge, Action and Change in Communities: a Health Case Study from Gujarat, Western India.

Lessons learned from the participatory action research (PAR) project helped identify and categorize four main areas in which practical challenges may arise, impeding abilities to transfer knowledge to action and thus change for communities (Read 2012).

The chapter commences with a review of the factors affecting participatory translation of knowledge to action in action-oriented projects, and the few published examples of responses to these challenges. The main focus of the chapter will be presenting the new tool to help us combat these challenges even before they arise. The Action Preparedness Tool for Community-based Projects (APTCP) has been formulated as an instrument to assist researchers and practitioners of participatory approaches to research and action-oriented projects to identify and categorize context-specific challenges that may arise during efforts to implement participatory action with communities. Applying APTCP prior to commencing a project and asking a series of vital questions can help prepare us for community-based action and avoid common pitfalls throughout the duration of the project.

What Experience Tells Us

Expanding on the experience of Chap. 6 and creation of Fig. 6.6, Fig. 14.1 offers a more complete view of the process in question.

Figure 14.1 is a visual representation of the factors affecting the participatory translation of knowledge to action in community-based, action-oriented research/projects. At the top of the diagram is the cyclic and reciprocal relationship between knowledge, action and change. These are the expected outputs of the research/project. The relationship in focus is the translation of knowledge to action: the bolded arrow between the two outputs. Numerous factors affect this relationship and these are numbered 1–4 in the circular diagram.

They include factors relating to the:

1. Research environment
2. Organization
3. Community
4. Researcher/Practitioner

Awareness of these factors affecting the transition from knowledge to action is the first and most important step towards addressing the challenges inherent in action-oriented projects. Awareness of these factors must then be followed with response.

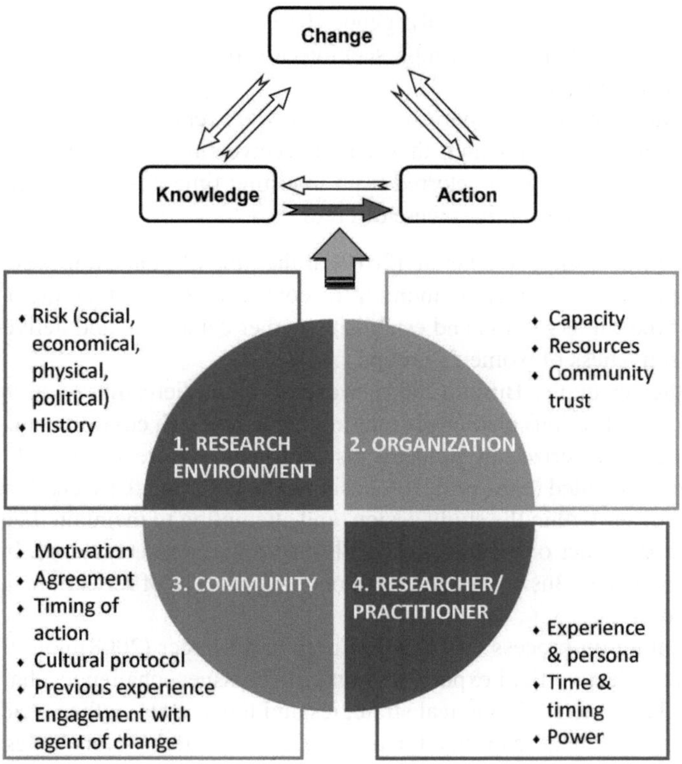

Fig. 14.1 Factors affecting participatory translation of knowledge to action in action-oriented projects

Response to Challenges Inherent in Action-Oriented Research Projects

The few existing examples of literature discussing the challenges researchers experienced throughout their action-oriented project, and how they responded to the challenges, outline reactive processes. The problems encountered, much like the example in Chap. 6, were only realised once the project had commenced, giving little opportunity to overcome the difficulties in feasible time.

Azad et al. (2010), for example, in their assessment of the effectiveness of a scaled-up development programme with women's groups to address maternal and neonatal care in three rural districts of Bangladesh, found that participatory women's groups did not significantly reduce neonatal mortality in poor rural populations of Bangladesh. They call for detailed attention to contextual factors that affect women's participation in action-focused groups. Correlating to the categories of challenges identified in Chap. 6 and expanded upon in Fig. 14.1, the main contextual challenges experienced include:

- **Researcher/Practitioner** (time and timing): adverse climatic conditions affected the facilitators' ability to travel to meetings;

- **Community** (cultural protocol): gender-based barriers were strong preventing some women from joining groups, seeking care, or from implementing strategies if they had joined a group;
- **Community** (previous experience): women were regularly asked to participate in NGO activities for which they could receive financial incentive, therefore women might have been deterred from investing time in women's group meetings for which no incentives were offered.

To address these issues, Azad et al. (2010) at the time of publishing were applying a detailed process assessment to monitor the contextual factors that might affect the delivery of their intervention and establish whether contextual and delivery factors reduce effectiveness of women's groups.

In another example, Busza (2004), working with Vietnamese sex workers in Cambodia, faced serious challenges relating to the research environment. The local context was characterized by political instability and severe social and legal constraints which resulted in the project focusing on negotiating these challenges, making compromises within the study design, and attempting to maintain their commitment to the principles of participation while ensuring rigorous research. In response to these challenges, Busza engaged in an ongoing process of reflection, adaptation, and compromise.

Commenting on process evaluation, Cargo and Mercer (2008) note that evaluation of the reflective field experiences on critical issues, challenges, barriers, and facilitators has generated practical strategies and tools to strengthen academic and non-academic partners' capacity for participatory research. 'Nonetheless, lack of consistency in the use and measurement of core process indicators and the lack of comparative case studies have limited progress in understanding how variation in implementing participatory research approaches relates to research outcomes' (Cargo and Mercer 2008, p. 340).

Another issue is that process evaluation may assist researchers to identify the challenges, but only once they occur. In that case, response is reactive. What if there was a way these challenges could be identified before they hinder the project, and researchers would have the opportunity to put in place proactive measures to overcome challenges? The APTCP presented in the following section aims to fill that role and offer researchers and practitioners a new approach to community development and participatory research.

Other examples of efforts to proactively tackle challenges of community-based, action-oriented research do exist. Ross et al. (2010) provide a 'points-to-consider' roadmap for academic and community research partners to establish and maintain a research partnership at each stage of the research process. Green et al. (1995) offer review criteria for community-based participatory research proposals and a rating scale to these criteria to help researchers and their partners discern the degree to which their project design is participatory and action-oriented. The tool introduced in this chapter differs, in that it categorizes key questions under the categories of key project stakeholders and focuses the questions specifically to elicit information that will enable self-assessment of preparedness for action. The following section expands on this concept and provides examples of literature that justify why the questions in the tool must be asked.

Action Preparedness Tool for Community-Based Projects

This tool has been formulated as an instrument to identify and categorize context-specific challenges that may arise during efforts to implement participatory action with communities. This tool can be applied by any or all stakeholders of the community-based action-oriented project, including the community themselves, researcher/practitioner, organization and funding body. Singh et al. in Chap. 3 acknowledge that community development, whatever the context, will face potential barriers that could block individuals' participation. They suggest that barriers should be addressed by the community as a whole. Following this train of thought, it is recommended for the tool to be applied as a collective process, including communities and other stakeholders in the process.

This tool is intended to be applied prior to commencing an action-oriented project and once community issues/problems are identified. It can highlight deficiencies and enable stakeholders to measure the likelihood of efforts achieving participatory action. It can also indicate the likelihood of achieving participatory action to address community needs. This information will help identify the risk of projects failing due to contextual challenges. When this risk is addressed, participants can save valuable time, resources, energy and social capital with communities that may potentially be damaged by implementing a project that has a high likelihood of failing. The tool's relevance and scope reach beyond the health field to other disciplines working with communities to create action leading to change in communities, and it offers a new perspective to sustainable community development (Fig. 14.2).

The following sections provide a description of the necessary considerations to ensure stakeholders are prepared for action and the potential challenges that could arise during implementation.

Pre-Action Phase Considerations

Before we ask questions relating to the four main contextual factors (Research Environment, Organization, Community and Researcher/Practitioner), we must first ask a series of questions that relate to actions taken to arrive at this stage of the project. At this stage, when the tool should be applied, the community problem/issue has already been identified and stakeholders are poised, ready to commence action to address the problem. We must however be mindful that decisions already taken, such as how the issue was identified and who the key initiators of the project are, along with whether care has been taken to balance expectations and benefits and implement appropriate methodologies, will affect project outcomes. The following sections explain in more detail.

PRE-ACTION PHASE CONSIDERATIONS

- How was the issue/topic selected?
- Who are the initiators of change?
- Are the expectations of all stakeholders balanced?
- What is the chosen methodology/methods to be applied?
- Are the expected benefits of the research balanced?
- How has 'community' been defined?

RESEARCH ENVIRONMENT

- Is there risk involved in participating in the action-intervention?
- Does the community have any previous experience taking action on the problem being addressed? If so, were there any events that occurred that may affect current efforts?

ORGANIZATION

- Do the organization/initiators of change have the appropriate level of resources and capacity to implement the project?
- Do the organization/initiators of change have established trust with the community?

COMMUNITY

- Is the community currently engaged with the agent/initiator of change?
- Is the timing of participatory action appropriate (both time of year and matched with catalyst for change)?
- Is the community motivated for change?
- Does the community agree on the issue to be addressed?
- Has community specific cultural protocol been adhered to?
- Does the community have previous experience with participatory action research or related community-based, action-oriented methodologies?

RESEARCHER/ PRACTITIONER

- Does the researcher/practitioner have experience as well as personal qualities required for participatory research?
- Has the researcher/practitioner considered the effect of power imbalance between the researcher and participant/community?
- Has the researcher/practitioner allocated an appropriate amount of time for the project?

Fig. 14.2 Action preparedness tool for community-based projects

From the start, how the issue/topic was selected is of critical importance in shaping the process and outcome of the project. In research into 20 community-based participatory research (CBPR) projects, Cook (2008) found that CBPR projects that led to action are much more likely to have been initiated by the affected community or by government scientists/agencies than by academic researchers. They note that the needs of the researcher are often guided by theoretical research questions, whereas the needs of the community are practical and action-oriented. It is therefore not surprising that:

> … when the impetus of CBPR came from the affected community, the research problem was more likely to reflect genuine community concerns and the research to be carried out with potential, relatively near-term solutions in mind and to lead to meaningful action. (Cook 2008, p. 674)

Thus, the first question that must be asked is: *how was the issue/topic selected; and who are the initiators of change?* These simple questions can help us orient ourselves with the notion that community-led action will be more likely to achieve results. Should the project not be community-led, we must be prepared, from the start, for the possibility of a lesser outcome in terms of action and change.

An imbalance of expectations can greatly hinder a project. These may include the expectations of each partner in terms of their involvement and the level of action and change expected from the project. Often the expectations of community partners' participation may be overestimated. Israel et al. (2005) note that community partners, many of whom are not paid by the project and have numerous other professional responsibilities, may not be supported by their supervisors if their involvement is perceived to be taking time away from other organizational responsibilities, causing tension between the stakeholders. Also, the extent and level of the expected change must be balanced. Whilst researchers may place emphasis on the opportunity to affect the lives of large numbers of people through policy impact, in some instances community participants' desires for action may be limited to securing small, individual level changes. For the outside researcher, an ethical dilemma may therefore involve the tension between respecting community wishes and pushing for the higher-level social change that some consider a defining feature of CBPR (Minkler 2004). There are many areas where community and academic partners may differ in their emphasis on goals, values, priorities, and perspectives (Israel et al. 1998). For example, in another project, Israel et al. (2005) found that several community partners were eager to implement the interventions and disseminate preliminary results, whereas researchers were concerned that the premature dissemination of results would contaminate study findings and lead to scientific criticism and consequences for publications and future funding. The necessity of balancing expectations from the outset to avoid conflict between partners is therefore highlighted, and we must ask: *are the expectations of all stakeholders balanced?*

The methods and methodologies applied during the project will also have impact on project outcomes. We must then ask ourselves: *what are the chosen methodology/methods to be applied? Are they appropriate for the context?* In Cook's (2008) study of 20 action-oriented projects, they found that projects that integrated research with action are more likely to be observational studies, most of which incorporated qualitative methods. The qualities of qualitative methods render them particularly useful for CBPR. They are likely to help generate actionable knowledge, which has been informed by the community, and integrate it into contextually relevant strategies to promote community-level changes. Whilst qualitative methods are preferable for eliciting in-depth information from participants without the structured research protocols, we must consider if our methods are appropriate for the context. In the example of Busza (2004), the use of traditional data collection methods utilised for the HIV-prevention programme for Vietnamese sex workers in Cambodia received opposition. First, brothel owners were suspicious of wanting to speak to women privately for in-depth interviews; second, the sex workers felt that giving information on a one-to-one basis over a questionnaire was boring or a waste of time. In another study with an indigenous community in Canada, community members initially strongly objected to the idea of using a questionnaire approach which they saw as 'putting their thoughts in boxes' (Minkler 2005, p. ii9). These examples highlight the importance of considering the appropriateness of the methods used to invoke action in communities.

In terms of the benefits of the research, Minkler (2005) agrees that the major aim of CBPR is to benefit the local community; however benefits are not always spread, 'with outside researchers typically standing to gain the most from such collaborations' (Minkler 2005, p. ii9). When inequity among partners receiving benefits occurs, relationships can be strained and threaten their long-term viability, especially when resources are reduced (Israel et al. 2006). For this reason it becomes necessary to ask, *are the expected benefits of the research balanced?*

Another critical consideration in establishing a CBPR partnership is deciding *how the community is defined, who represents the community, and how partners are selected* (Israel et al. 2003). Where people have little sense of belonging to a community, they may have little inclination to spend time on community affairs (Cornwall 2008). How community has been identified therefore becomes a necessary question in the category of pre-action phase considerations. It is not enough to assume that people in a defined geographic area make a community, and we must place more emphasis on knowing who the community we will be working with is, and why it is a community.

The questions raised in this section on the pre-action phase considerations are designed to encourage stakeholders to think more deeply about the decisions being taken to implement an action-oriented project. Should it occur that: the project is being undertaken because of an organizational requirement rather than a community-led desire; one partner is set to benefit more than the other; expectations on what the project can practically achieve are skewed; or little thought has gone into defining or knowing the community; we must be prepared for outcomes to be hindered, even before contextual factors come into play.

The Research Environment

The research environment is the first of four contextual factors that may cause challenges during an action-oriented project. It is the responsibility of the practitioner/

researcher to assess the research environment and ensure that individuals and their communities are not put at risk by participating in a project that acts on a problem in their community. However, it is a recognized challenge of CBPR that 'participation in the action phase of CBPR projects may sometimes present risks to community participants and actions that involve challenging powerful corporate or other entrenched interests may have negative consequences for those involved' (Minkler 2005, p. ii10). Farquhar and Wing (2003, p. 238) point out that 'no research question, method or result can be separated from its social context'. As a result, both outside researchers and community participants must be prepared for the sometimes difficult consequences that this may entail (Minkler 2004). Risk can also be economic in nature. Community partner organizations face financial costs from involvement, such as lack of adequate reimbursement for their time spent participating, as well as opportunity costs for time taken away from other job responsibilities (Koné et al. 2000; Parker et al. 2003). When this risk of cost is elevated, participating in action may be avoided by community partners.

It is evident that projects that are likely to introduce an element of risk to participants and partners are less likely to result in both participation and action, therefore we must ask, *is there risk involved in participating in the action-intervention?*

Cornwall (2008) affirms that even the most transformational intentions can meet a dead end when intended beneficiaries choose not to take part. When a community has negative past experiences taking action on specific community problems, their involvement in the project can be greatly impeded. We must therefore ask: *does the community have any previous experience taking action on the problem being addressed? If so, were there any events that occurred that may affect current efforts?*

As an example, Cohen and Uphoff (1980) in their discussion of the task environment in which a project operates, identify such previous experience (historical factors) as one factor at the rural-agricultural level that can impact a project. They cite that prior experience with a project whose rice seeds failed to germinate may make it difficult to get new high yielding varieties adopted. Also (a possibility for a range of disciplines) history of embezzlement of self-help funds raised by the community may lead many local people to distrust a project.

Additionally, when communities have the past experience of being continually consulted and see nothing as a result, not participating 'may be a pragmatic choice to avoid wasting time once again' (Cornwall 2008, p. 280). 'Participation fatigue' (Cornwall 2008, p. 280) accounts for more and more active self-exclusion in action projects, and this problem, caused by negative past experience, should be taken into serious consideration.

Khanlou (2005, p. 2337) reinforces the importance of understanding the research environment, stating, 'it is imperative that an acute sensitivity to the politics and culture of communities is demonstrated before a participatory action research project is initiated'.

The research environment is, however, only one of the four main categories of contextual factors affecting knowledge to action transition. Factors relating to the organization, community and researcher/practitioner are just as important. The following sections will consider further questions to be asked as part of the Action Preparedness Tool for Community-Based Projects.

The Organization

This section of the tool relates to the lead organization involved in the community-based project. There is very little guidance on the institutional capabilities necessary to support and sustain CBPR and other participatory action-oriented projects, and discussion regarding the topic remains fragmented (Faridi et al. 2007). There are several necessary capabilities, for example: 'time, energy, resources, funding mechanisms, tenure structures, organizational hierarchy, research focus, power-sharing arrangements, and institutional commitment required to conduct CBPR and maintain successful partnerships with communities' (Faridi et al. 2007, p. 3). However, this tool highlights the importance of resources, capacity and community trust as the core necessities to carry out an action-oriented project.

The first question to be asked is: *do the organization/initiators of change have the appropriate level of resources and capacity to implement the project?* Whitmore (1998) agrees, and lists several questions that organizations and individual researchers should consider before embarking on a CBPR project, including whether the research team has the necessary skills to conduct the project, and whether the institution has the requisite resources and infrastructure to engage in this type of research. Costs for both community and academic partners to conduct participatory action-oriented projects are numerous, and as suggested, there are insufficient resources for overcoming them (Israel et al. 1998; Koné et al. 2000; Minkler 2004). An effective partnership requires time and infrastructure support in order to jointly participate in all phases of the research, foster capacity building and, of utmost importance, to establish and maintain trust (Israel et al. 2005).

Thus we arrive at the next crucial question: *have the organization/initiators of change established trust with the community?* Time, resources and capacity to build and maintain trust should not be underestimated. Building and maintaining trust between the university and community as well as at times among community partners is recognized as a substantial challenge (Israel et al. 1998; Minkler 2004). Should trust have not been established prior to commencing a project, we must be prepared to invest significant time and resources to do so.

The Community

The community is of course a major stakeholder in community-based action projects, and questions related to their future involvement in creating action should be considered as important as the other three contextual factors—research environment, organization and researcher/practitioner. Related to the previous question, whether the organization/initiators of change have established trust with the community, is the question: *is the community currently engaged with the agent/initiator of change?* Significant additional time would be required if the community does not yet have an established relationship with the agent of change. In Chap. 5, for instance, Madigan and Morris had a very short 2 week time frame to conduct their project. They entered the community with an understanding of the importance of developing a strong relationship with the community. Their project outcomes were positive but should they have received resistance from the community, their entire project could have been jeopardized.

Timing is of great importance to conducting an action-oriented project, and can sometimes be overlooked in attempts to schedule activities into academic or funding calendars: *is the timing of participatory action appropriate (both time of year and matched with catalyst for change)?* In many rural areas for example, long rainy seasons may make it impossible to hold regular co-operative meetings throughout the year because roads and paths are impassable (Cohen and Uphoff 1980). Adverse climatic conditions affecting the facilitators' ability to travel to meetings was cited as a contextual factor faced in the study of Azad et al. (2010), and whilst it may seem logical that seasons would be accounted for, it is not uncommon to find that little thought goes into the timing and duration of participatory activities (Cornwall 2008). This also includes requesting the involvement of people who work, people who have small children to put to bed or feed, people who are unable to justify spending hours outside the household (Cornwall 2008).

An important finding by Cook (2008) is that action-oriented projects can be more successful when supported by social movements. They found numerous examples of CBPR in which community partners who had built their organizational strength in related social movements, such as environmental justice and labour movements, came to play an integral role in CBPR to generate tangible benefits for the community.

Many times, an organization's mandate, funding requirements, or academic hypotheses dictate the issue to be addressed in a community. However, when the topic for a potential research partnership does come from an outsider, Minkler (2004) highlights that a key preliminary step should involve determining whether the proposed subject really is high on the agenda of the affected community. *Is the community motivated for change?* is a question that should not be overlooked. For CBPR to truly energize and empower communities, it is essential for community members to have a critical consciousness and increased desire and ability to mobilize their community to effect change (Cook 2008). To conduct a successful participatory action-oriented project, the community must therefore be mature enough to offer a vision and have the ability to jointly investigate community issues and plan a course of action to address them (Cook 2008).

Motivation may be key, but we also have to consider: *does the community agree on the issue to be addressed?* CBPR is committed to strengthening community capacity and trust, so choosing an issue or topic that may further divide community members may hold considerable risk (Minkler 2004).

With a topic of importance decided upon, the next question is: *has community-specific cultural protocol been adhered to?* Gaining consent, either formally or informally, from a community leader or elder before implementing a study is an important consideration in developing countries, particularly rural areas (Marshall 2006, 2007). In a survey of more than 500 researchers from developing countries, Kass and Hyder (2001, as cited in Marshall 2007) found that more than half sought approval from a community leader or village elder for documenting consent or informing participants about their study and access to a site. Specific cultural protocol will differ from context to context. In areas where there is high gender inequality, it may not be uncommon that gender-based barriers prevent some women from joining groups, seeking care, or from implementing strategies if they have joined a group; see, for example, Azad et al. (2010). Azad also found that some women had to ask permission from their husbands or in-laws to join a group, which was problematic.

The question, *does the community have previous experience with participatory action research or related community-based, action-oriented methodologies*, has implications whether the answer is yes or no. Experienced communities may have experienced so many attempts to participate that they have become tired and cynical (Cornwall 2008). It is difficult to 'sell' the concept of participatory action-oriented research to community members, especially when the expected benefits aren't tangible. 'You can't eat participation, can you?' was the response of one of Cornwall's (2008, p. 274) disgruntled would-be community representatives, who was expected to mobilize his fellow community members without receiving any of the handsome salaries that his NGO counterparts were paid. Certainly when other organizations are operating in the region and they offer financial incentive for participation, community participants might be deterred from investing time in group meetings for which no incentives are offered (Azad et al. 2010).

The Researcher/Practitioner

Applying participatory approaches to action-oriented research projects such as PAR and CBPR requires not only an understanding of the orientation of research, but the correct personal attitude and attributes. As a research methodology separate from conventional approaches to research, PAR requires a very different set of skills, including self reflection and reflexiveness (Rice and Ezzy 1999, p. 188). The effectiveness of the PAR process depends on a researcher's abilities to negotiate with a community (Westby 2003, p. 304). Westby (2003, pp. 302–304) proposes a number of considerations and recommendations to guide researchers and practitioners in initiating and maintaining PAR when working cross-culturally. They include: cultural sensitivity/awareness; ability to build collaborative relationships and trust; respect for and adherence to cultural protocols; interpersonal skills including empathy and openness; and willingness to try out alternative constructs to explain the personal worldviews. It becomes necessary to ask ourselves: *does the researcher/practitioner (do I) have experience as well as personal qualities required for participatory research?* Israel (2005) identifies limited training and experience in conducting CBPR as a challenge, despite the large and growing literature on how to carry out CBPR efforts.

Further, issues of power imbalance and insider-outsider tensions related to privilege and control are widely discussed in the literature (Fadem et al. 2003, p. 254; Israel et al. 1998, p. 183; Minkler 2004, p. 688; Minkler 2005, p. ii8; Wallerstein and Duran 2006, p. 313). 'Even outsiders who pride themselves on being community allies and trusted friends frequently fail to realize the extent of the power imbued by their own, often multiple, sources of privilege and how it can adversely affect interactions and outcomes' (Fadem et al. 2003, p. 254). Distribution of information, time, formal education and income reflect broader social inequities structured around race/ethnicity, class and gender, and these affect who participates, whose opinions are considered valid, and who has influence over decisions made in a community based research partnership (Israel et al. 1998, p. 183). Thus the question arises: *has the researcher/practitioner considered the effect of power imbalance between the researcher and participant/community?* Authors of Chap. 2, Tiwari and Pandya, experienced an unexpected challenge which resulted in the power residing with one stakeholder. In their case, the NGO held the responsibility of the direct link to the community, researchers, architect and the government. Unintentionally omitting local government bodies had unforeseen consequences.

The time-consuming nature of CBPR is continually reiterated in literature, and is viewed as 'one simple manifestation of insider-outsider tensions that may involve conducting a participatory community project within the timeframe of the academic and funding calendar' (Fadem et al. 2003, p. 254). We must therefore specifically ask: *has the researcher/practitioner allocated an appropriate amount of time for the project?* Real collaboration takes a lot of time, for meetings, for accountability processes, for working through the inevitable conflicts (Stoeker 2003, p. 101) and to establish and maintain trusting relationships (Israel et al. 1998, p. 182). In Karmaliani et al.'s CBPR study to improve maternal and child health in Karachi, Pakistan (2009), one year was spent building a coalition with the community before even applying for research funding. It is true that researchers are often in an impossible situation when it comes to the promotion of participatory methods, because time is often squeezed and does not permit the required time for dialogue (Mikkelsen 1995). We must however be mindful that CBPR is a long term commitment (Holkup et al. 2004; Rice and Ezzy 1999).

The processes of CBPR take investment, time and persistence, and as eloquently noted by Cornwall 'they cannot be achieved by waving a magic participation wand, convening a participatory workshop or applying a few Participatory Rural Appraisal tools and hey presto, there is empowerment!' (Cornwall 2008, p. 278).

The Way Forward

Too often, action-oriented research and projects fail to create sustainable change for communities facing hardship. For the sake of communities, these failures must be scrutinised, and reasons why interventions have not achieved desired outcomes must be documented. This chapter argues that contextual factors relating to the research environment, community, researcher perceptions and organizational capacity all have a role in affecting the transformation of knowledge to action, and essentially the success or failure of an intervention (in terms of ability to influence positive change). These factors are recommended to be applied as an Action

Preparedness Tool for Community-Based Projects before commencing action. This will give researchers an indication of factors requiring amendment to meet the set requirements, and given the current situation, the chances of transforming knowledge to action.

Tested methodologies, approaches and methods alone cannot ensure a successful outcome to the transition from knowledge to action. External contextual factors that are separate from methodological decisions impact on participatory action, and combined with the complex nature of community problems, can cause less desirable outcomes. The Action Preparedness Tool for Community-Based Projects proposed in this chapter can provide solutions to this problem, and it is proposed that initial project resources are better directed by making this assessment of whether community efforts are likely to evoke action that leads to beneficial and sustained change. Although participatory approaches to research such as PAR and CBPR are inherently challenging, particularly when applied in disadvantaged communities in complex environments, there is hope that with continual improvements, community-led action research can bring about change for communities.

Acknowledgements This study was undertaken as a collaborative community project between the Centre for International Health (CIH) at Curtin University, Western Australia and the Centre for Rural Development (CRD), SNDT Women's University in Mumbai, India.

Funding for this study was provided by the Centre for Advanced Studies in Asia, Australia and the Pacific (CASAAP) and in-kind support was provided by the CRD.

The primary researcher expresses gratitude to her PhD supervisors, Associate Professor Jaya Earnest and Mohammed Ali from CIH, Curtin University for their support throughout the entire research journey. Also to all employees at the Center for Rural Development, Udwada, who assisted with this project, including Director of the Research Center for Women's Studies, Professor Veena Poonacha at SNDT University. A special acknowledgement is made to Suman Bali, past Coordinator, Health Project, whose support during the research was invaluable.

This study could not have been possible without the participation of the communities of the five villages in Southern Gujarat and we sincerely thank those who participated and gave their time.

References

Azad, K., Barnett, S., Banerjee, B., Shaha, S., Khan, K., Rego, A. R., et al. (2010). Effect of scaling up women's groups on birth outcomes in three rural districts in Bangladesh: A cluster-randomised controlled trial. *The Lancet, 375,* 1193–1202. doi:10.1016/S0140-6736(10)60142-0.

Busza, J. (2004). Participatory research in constrained settings sharing challenges from Cambodia. *Action Research, 2*(2), 191–208.

Cargo, M., & Mercer, S. L. (2008). The value and challenges of participatory research: Strengthening its practice. *Annual Review of Public Health, 29,* 325–350.

Cohen, J., & Uphoff, N. (1980). Participation's place in rural development: Seeking clarity through specificity. *World Development, 8,* 213–235.

Cook, W. K. (2008). Integrating research and action: A systematic review of community-based participatory research to address health disparities in environmental and occupational health in the USA. *Journal of Epidemiology and Community Health, 62*(8), 668–676.

Cornwall, A. (2008). Unpacking 'participation': Models, meanings and practices. *Community Development Journal, 43*(3), 269–283. doi:10.1093/cdj/bsn010.

Fadem, P., Minkler, M., Perry, M., Blum, K., Moore, L., & Rogers, J. (2003). Ethical challenges in community based participatory research. In M. Minkler & N. Wallerstein (Eds.), *Community based participatory research for health* (pp. 242–262). San Francisco: Jossey-Bass.

Faridi, Z., Grunbaum, J. A., Gray, B. S., Franks, A., & Simones, E. (2007). Community-based participatory research: Necessary next steps. *Preventing Chronic Disease. Public Health Research, Practice, and Policy*, *4*(3), A70–74. http://www.ncbi.nlm.nih.gov/pmc/articles/PMC1955426/. Accessed 21 Apr 2012.

Farquhar, S., & Wing, S. (2003). Methodological and ethical considerations in community-driven environmental justice research. In M. Minkler & N. Wallerstein (Eds.), *Community-based participatory research for health* (pp. 221–241). San Francisco: Jossey-Bass.

Green, L. W., George, M. A., Daniel, M., Frankish, C. J., Herbert, C., Bowie, W. R., et al. (1995). Study of participatory research in health promotion. *Review and recommendations for development of participatory research in health promotion in Canada.* Ottowa: The Royal Society of Canada.

Holkup, P. A., Tripp-Reimer, T., Salois, E. M., & Weinert, C. (2004). Community-based participatory research: An approach to intervention research with a Native American community. *Advances in Nursing Science*, *27*(3), 162–175. http://www.ncbi.nlm.nih.gov/pmc/articles/PMC2774214/. Accessed 21 Apr 2012.

Israel, B., Schulz, A. J., Parker, E. A., & Becker, A. B. (1998). Review of community-based research: Assessing partnership approaches to improve public health. *Annual Review of Public Health, 19,* 173–202. doi:10.1146/annurev.publhealth.19.1.173.

Israel, B., Schulz, A. J., Parker, E. A., Becker, A. B., Allen III, A. J., & Guzman, J. R. (2003). Critical issues in developing and following community based participatory research principles. In M. Minkler & N. Wallerstein (Eds.), *Community based participatory research for health* (pp. 53–76). San Francisco: Jossey-Bass.

Israel, B., Parker, E. A., Rowe, Z., Salvatore, A., Minkler, M., Lopez, J., et al. (2005). Community-based participatory research: Lessons learned from the Centers for children's environmental health and disease prevention research. *Environmental Health Perspectives, 113*(10), 1463–1471. doi:10.1289/ehp.7675.

Israel, B., Krieger, J., Vlahov, D., Ciske, S., Foley, M., Fortin, P., et al. (2006). Challenges and facilitating factors in sustaining community-based participatory research partnerships: Lessons learned from the Detroit, New York City and Seattle Urban research centers. *Journal of Urban Health, 83*(6), 1022–1040.

Karmaliani, R., McFarlane, J., Asad, N., Madhani, F., Hirani, S., Shehzad, S., et al. (2009). Applying community-based participatory research methods to improve maternal and child health in Karachi, Pakistan. *Nursing Outlook, 57*(4), 204–209. doi:10.1016/j.outlook.2009.01.003.

Khanlou, N., & Peter, E. (2005). Participatory action research: Considerations for ethical review. *Social Science & Medicine, 60,* 2333–2340. doi:10.1016/j.socscimed.2004.10.004.

Koné, A., Sullivan, M., Senturia, K. D., Chrisman, N. J., Ciske, S. J., & Krieger, J. W. (2000). Improving collaboration between researchers and communities. *Public Health Reports, 115*(2–3), 243.

Marshall, P. A. (2006). Informed consent in international health research. *Journal of Empirical Research on Human Research Ethics, 1*(1), 25–42. doi:10.1525/jer.2006.1.1.25.

Marshall, P. A. (2007). Ethical challenges in study design and informed consent for health research in resource-poor settings. *Special topics in social, economic and behavioural (SEB) research report series No. 5.* World Health Organisation.

Mikkelsen, B. (1995). *Methods for development work and research.* London: Sage.

Minkler, M. (2004). Ethical challenges for the "outside" researcher in community based participatory research. *Health Education and Behavior, 31*(6), 684–697. doi:10.1177/1090198104269566.

Minkler, M. (2005). Community based research partnerships: Challenges and opportunities. *Journal of Urban Health: Bulletin of the New York Academy of Medicine*, *82*(2), Sup ii3–12. doi:10.1093/jurban/jti034.

Parker, E. A., Israel, B. A., Williams, M., Brakefield-Caldwell, W., Lewis, T. C., Robins, T., et al. (2003). Community action against asthma. *Journal of General Internal Medicine, 18*(7), 558–567.

Read, C. (2012). *Applying a participatory action research model to assess and address community health concerns among tribal communities in Gujarat, Western India: The potential and challenges of participatory approaches*. Doctoral thesis. Curtin University, Published online on the Australian Digital Thesis Website. http://espace.library.curtin.edu.au:80/R?func=dbin-jump-full&local_base=gen01-era02&object_id=188776. Accessed 21 April 2013.

Rice, P. L., & Ezzy, D. (1999). *Qualitative research methods: A health focus*. Melbourne: Oxford University Press.

Ross, L. F., Loup, A., Nelson, R. M., Botkin, J. R., & Kost, R., Smith Jr, G. R., et al. (2010). The challenges of collaboration for academic and community partners in a research partnership: Points to consider. *Journal of empirical research on human research ethics: JERHRE, 5*(1), 19.

Stoeker, R. (2003). Are academics irrelevant? Approaches and roles for scholars in community-based participatory research. In M. Minkler & N. Wallerstein (Eds.), *Community based participatory research for health* (pp. 98–112). San Francisco: Jossey-Bass.

Wallerstein, N., & Duran, B. (2006). Using community-based participatory research to address health disparities. *Health Promotion Practice, 7*(3), 312–323. doi:10.1177/1524839906289376.

Westby, C., & Hwa-Froelich, D. (2003). Considerations in participatory action research when working cross-culturally. *Folia Phoniatrica et Logopaedica, 55*(6), 300–305. doi:10.1159/000073253.

Whitmore, E. (1998). *Understanding and practicing participatory evaluation: New directions for evaluation*. San Francisco: Jossey-Bass.

Part III
Community Engagement and Capacity Building: a Transdisciplinary Perspective

This final part of M2 takes a more personal position as it reflects on the understandings that are evolving for the editors. As members of the book's collective, we draw upon the issues raised and the lessons learnt and shared by the other contributors. We posit that ways of conceptualizing key terms and practices may be of value when we address the evolution of interactions on a global scale. Personal connectedness is increasing, and increasingly complexified, through innovations such as digital media, combined with social issues such as homelessness and mental illness, and environmental issues such as climate change and extreme natural weather patterns. All these challenges demand that we attend to life quality for all humans. To do so, an awareness of how the practice of community engagement is enacted, and participation defined and controlled, is required. This section asks the reader to consider the taken-for-granted attributes and those that are still evolving in the light of the contextual influences we outline, in order to develop appropriate and empathetic practices and processes. How community is understood and by whom is an important topic in these discussions.

An election is coming. Universal peace is declared and the foxes have a sincere interest in prolonging the lives of the poultry.

George Eliot, *Felix Holt, the Radical*

Chapter 15
Transforming Community: Opportunities for Transdisciplinary Application

Marina Lommerse, Dianne Smith and Reena Tiwari

Abstract This chapter discusses how individual authors in Parts I and II of this book have understood and investigated the idea of community. We look at whether and how their definition of community is dependent on location and the nature of their projects, how their views changed through reflection and interaction with community members, and what constitutes capacity building in different socio-cultural contexts. The projects discussed capture a multitude of ways to interface and interact with communities, of modes of interaction, empowerment or engagement, of varying scales and complexity and socio-economic circumstances, as well as of physical locations across the world. The transformative experience is valid, not only for local users/residents in the project, but also for the associated civil society, the policymakers, the professionals, the students and other social actors involved. The case studies described in the chapters in Parts I and II are reflected on in terms of the concept of community as a discipline in its own right, which builds capacity in a transformative experience that brings physical, social, economic or environmental change. We revisit notions of co-creation, capacity building and transformation, and identify transdisciplinary understandings of community engagement. The editors have arrived at the notion of community as a discipline and the increased value of recognizing community engagement as a transdisciplinary practice.

Keywords Capacity building · Transformation · Community development · Transdisciplinary community engagement · Co-design · Co-production

---

M. Lommerse (✉)
Marina Lommerse Consultants, South Fremantle, Australia
e-mail: marina@marinalommerse.com

School of the Built Environment, Curtin University, Perth, Australia

D. Smith
Department of Architecture and Interior Architecture, Curtin University, Perth, Australia
e-mail: dianne.smith@curtin.edu.au

R. Tiwari
Departments of Urban and Regional Planning, and of Architecture and Interior Architecture, Curtin University, Perth, Australia
e-mail: r.tiwari@curtin.edu.au

R. Tiwari et al. (eds.), *M² Models and Methodologies for Community Engagement*,
DOI 10.1007/978-981-4585-11-8_15,

Introduction

This concluding chapter offers the editors' insights into the wider implications that can be drawn from the different voices in this book to contribute to a developing, transdisciplinary understanding of community engagement. We find that transdisciplinary community engagement partnerships/projects have already gained momentum, and will continue to do so. Hence there is a demand for best practice models and methods that rationalize the complex and diverse territory to create a space where academics, professionals, nonprofit institutions and community can merge, sharing knowledge-action opportunities that engage all. To help build a framework for such shared creative endeavours, this chapter expands on the editors' positioning of community as a discipline, as argued in Chap. 1, and on the complex and ill-defined concept of transdisciplinarity and its implications for community engagement and capacity building.

This book began by identifying the following questions:

- How do researchers and professionals engage with the community? What is the impact on the form and degree of engagement of different disciplinary perspectives, the range of attributes, various scales, different locations and whether the project is set in a developing or developed world context?
- Does this community engagement assist to build the capacity of the local community members and empower them personally, socially or economically?
- If so, what are the models and methods required for successful or meaningful engagement? What tools are integrated to facilitate the necessary interactions with community?

As we responded to these questions, firstly by using visual techniques in Chap. 11 to describe Part I models, the editors have examined the methodologies and theories described, the tools used, the context of the communities, and the concepts of power sharing and participation discussed by contributing authors. We have considered commonalities and diversity in community engagement from the included discipline perspectives, and the significance of what constitutes capacity building in different socio-cultural contexts. Building on Chap. 11, in this chapter the editors continue the investigation and fold in further narratives from Part II, to consider the barriers, challenges and most importantly, the potential of community engagement as we define it.

In Chap. 1, we hypothesized the concept of community as a discipline in its own right, which builds capacity in a transformative experience that brings physical, social, economic or environmental change. This chapter revisits the notion of community through a transdisciplinary lens, which we have used to investigate all the contributing authors' projects. The outcome of this approach has been the new concept of community as a discipline. Hence the second section of this chapter investigates the idea of community as understood by contributing authors, and then summarizes the barriers, challenges and potential of community engagement. We follow with the implications this has for framing community as a discipline in the context of transdisciplinarity. We argue that projects done with an understanding of community as a discipline can lead to transformation for all involved. Thus the third section explores capacity building as a transformative process. We conclude by reprising the new concept of community that has emerged from this book.

Transdisciplinarity in Community Based Projects

When we examine the common barriers, challenges and potentials in the different disciplinary approaches of the contributing authors, in their understanding of community and their ways of engaging, we open the door to other approaches. To set the groundwork for the following discussion, this section talks about useful concepts that have emerged in the process of writing the book and through the examination of contributing authors' models of engagement. We expand on the meaning, use and importance of transdisciplinarity in community-based projects, including related concepts, wicked problems, and co-creation/co-design/co-production.

The Emergence of Transdisciplinarity

Transdisciplinarity emerged in the 1970s (Hoffmann-Riem et al. 2008) as a response to a segregation of disciplines in the social sphere (Hoffmann-Riem et al. 2008; Russell et al. 2008; Novy 2012). Transdisciplinarity, like the term *sustainability*, has the potential to be ignored as a 'buzz' word, that is presented as a 'one-size-fits-all' concept (Russell et al. 2008). There is a plethora of definitions and constructs of the term transdisciplinarity, but an inclusive definition is that this methodology works across (*trans-*) discursive and cultural practices (*disciplinary*) to bring together structures, institutions and agencies (Miciukiewicz et al. 2012), in an effort to combine academic and non-academic knowledge and produce united knowledge (Doucet and Janssens 2011; Ramadier 2004). Ramadier (2004) importantly reminds us that new knowledge is not necessarily *produced*, but is rather a result of existing knowledge being articulated and, we would add, combined.

Tress et al. (2005) have defined interdisciplinary, multidisciplinary and transdisciplinary studies in terms of the degree and strength of integration. 'We define transdisciplinary studies as projects that both integrate academic researchers from different unrelated disciplines and non-academic participants, … to research a common goal and create new knowledge and theory' (Tress et al. 2005, p. 17). Figure 15.1, adapted from Tress et al.'s (2005, p. 17) diagrammatic illustration of integrative research models, illustrates the increasing degree of integration from participatory/multidisciplinary to interdisciplinary/transdisciplinary. Tress et al. stress that 'The main differences between the proposed concepts are the intensity of cooperation and integration of disciplines and the involvement of non-academic fields' (Tress et al. 2005, p. 17).

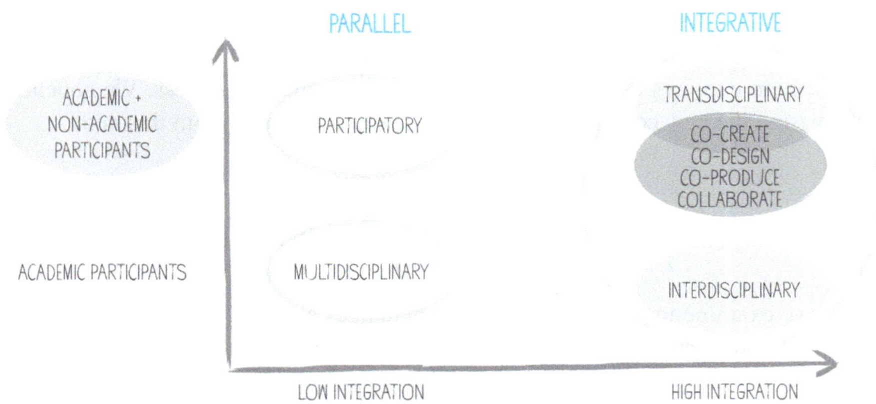

Fig. 15.1 High integration of various stakeholders (academic and non-academic), suggests an integrative, transdisciplinary approach, based on co-creation and collaboration. (Diagram adapted from Tress et al. 2005, p. 17)

Transdisciplinarity has emerged as a research practice that foregrounds the issue of research as the focus rather than discipline-specific methods and theories (Leavy 2011). Cassinari et al. (2011, p. 40) further defines transdisciplinarity as follows:

> [Transdisciplinarity]… recognises the dynamics of similarities across disciplinary knowledge (e.g. community development, social work, social planning). And disciplines here refer not only to scientific disciplines, but equally to practice fields also. [The goal of transdisciplinary research] is the holistic understanding of the world through the connections and unity of knowledge (Nicolescu 2002)…. Sometimes, a new 'discipline' may emerge, such as political ecology, cultural geography, complexity (physics, philosophy, cybernetics). Essential to transdisciplinary research is cooperation between scientists and those practitioners working outside academic communities.

As this book has developed in our community of writing, the editors' views have changed. At the book's conception we laid out Part I within a framework of communities of locale, communities of knowledge and communities of practice. Over 18 months, when we paused at chapter and book development milestones along the way, we facilitated writing and brainstorming workshops with contributing authors. These authors come from a variety of discipline areas, with different methodologies and conceptions of community. At the end, through the process of developing the ideas together in our writing and workshop discussions, we (the editors) have arrived at the notion of community as a discipline, and the increased value of recognizing community engagement as a transdisciplinary practice.

Transdiscplinarity can be thought of, in a graphic way, as rhizomatic thinking and practice, to adopt Deleuze and Guattari's useful distinction between arborescent and rhizomatic thinking (1987). 'Arborescent (tree-like) thinking is characterized by linearity, hierarchy, fixity, and deep rootedness' (Baldwin 2013, p. 106) as in, for instance, the traditional demarcations of academic disciplines, of methodologies in scientific/quantitative versus humanistic/qualitative research, and of social groups or classes in communities. In contrast, rhizomatic thinking is characterized

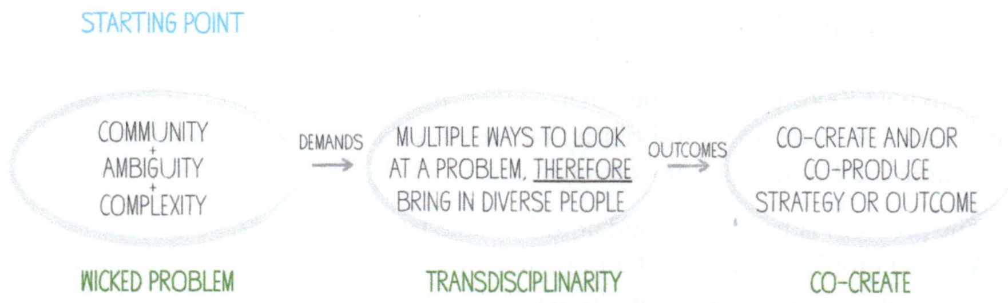

Fig. 15.2 Connections between wicked problems, transdisciplinarity and co-creation

by 'non-linearity, horizontal relations', and pathways that are 'non-centred, anarchic, and nomadic' (Baldwin 2013, p. 106). This allows for connections to be made throughout the rhizome (albeit, perhaps, indirect ones). 'We can enter the rhizome at any point, from which we can move to any other point' Baldwin 2013, p. 107), forming connections or relationships between those things that we assemble in constituting ourselves or our practices. Thus, in this chapter, we can trace connections, if not direct lines, between the different aspects of the projects our researchers have engaged in.

This also allows for distance in time and memory—the outreaches of the rhizome that are available to use but perhaps are rarely visited or even forgotten about until something prompts us to tread again that pathway (Baldwin 2013). This situates the spirit of community, not as a hierarchy, but as rhizomatic and holistic, meaning you can enter it at any point, and your entry point is less important than the connections within, which can lead to surprising new, transformative formations of change, taking communities to new possibilities of creativity and innovation. Here we are speaking about how the stakeholders look at the situation, rather than the project plan (or model) which may be more linear.

Transdisciplinarity needs to be flexible, responsive, sustainable and democratic (Novy 2012; Russell, Wickson and Carew 2008). Despite being complex, it is clear that the transdisciplinary approach is the key to holistic problem solving at a local level (O'Campo et al. 2011). Because it is system-oriented and not object-oriented, transdisciplinarity is a positive move toward better methods of capacity building and community engagement (Russell et al. 2008).

Next, we look at different aspects that are of relevance to transdisciplinarity: (a) wicked problems and (b) co-design and co-production. Figure 15.2 illustrates the connectedness between these problems and responses.

a) Wicked Problems

A wicked problem is one that is difficult to solve because of incomplete, contradictory, and changing requirements that are often difficult to recognize, and is characterized by its resistance to resolution. Rittel and Webber (1973) coined the term in the context of problems of social policy and planning; their formulation specified ten characteristics.

Conklin (2006, p. 43) argues that wicked problems can occur in any domain involving stakeholders with differing perspectives, and generalizes the concept to other areas, defining the characteristics as:

1. The problem is not understood until after the formulation of a solution.
2. Wicked problems have no stopping rule.
3. Solutions to wicked problems are not right or wrong.
4. Every wicked problem is essentially novel and unique.
5. Every solution to a wicked problem is a 'one shot operation.'
6. Wicked problems have no given alternative solutions.

A recurring theme in literature is the connection between wicked problems and design (Rittel 1988; Stolterman 2008). Design problems are typically wicked because they are often ill-defined (no prescribed way forward), involve stakeholders with different perspectives, and have no right or optimal solution (Conklin 2005). Thus wicked problems cannot be solved by the application of standard (or known) methods; they demand creative solutions (Conklin 2007).

Roberts (2000, pp. 4–6) identifies the following three strategies to cope with wicked problems: (a) authoritative, (b) competitive and (c) collaborative.

a. Authoritative strategies seek to tame wicked problems by vesting the responsibility and power for solving the problems in the hands of a few people. The reduction in the number of stakeholders reduces problem complexity, making it more expedient, as many competing points of view are eliminated at the start. The disadvantage is that authorities and experts charged with solving the problem may not have an appreciation of all the perspectives needed to tackle the problem.
b. Competitive strategies attempt to solve wicked problems by pitting opposing points of view against each other, requiring parties that hold these views to come up with their preferred solutions. The advantage of this approach is that different solutions can be weighed up against each other and the best one chosen. Central to competitive strategies is the search for power. The disadvantage is that this adversarial approach creates a confrontational environment in which knowledge sharing is discouraged.
c. Collaborative strategies are 'premised on the principle that by joining forces parties can accomplish more as a collective than they can achieve by acting as independent agents' (Roberts 2000, p. 6). Typically these approaches involve meetings in which issues and ideas are discussed and a common, agreed approach is formulated. The advantages of collaboration are numerous, including sharing the costs, risks and benefits. The most compelling is the premise that by joining forces (authorities, experts and locals) the approach is more likely to be right about the problem and right about the solution. The disadvantages of collaboration are that the problem-solving effort increases meetings, the number of people involved, and effort and time. Establishing operating procedures, and develop-

ing skills of collaboration, participation and team-based approaches to problem-solving and decision-making take additional time and resources.

Nonetheless, it is collaborative ones we are interested in here. These strategies aim to engage all stakeholders in order to find the best possible solution for all. Co-design and co-production are implicated here.

b) Co-design and Co-production

This is related to the knowledge production process introduced in Chap. 1—co-creation, co-design or co-production of knowledge, and/or systems and/or outcomes. The emphasis here is on bottom-up discourse formation, as opposed to traditional authoritarian, imposed definitions and parameters. Co-production aims for collaborative ways to create knowledge and address problems occurring in specific contexts (Fontan et al. 2013). It is a 'process by which the user [community] is led to participate in the production of the service [development, building] she/he requests' (Vaillancourt 2009; as cited in Fontan et al. 2013, p. 312).

Design processes are very suited to wicked problems, as design practices integrate all aspects of a situation; they consider, for instance, that people, processes and technology problems have to be considered equally, and at the same time.

Although transdisciplinarity is much discussed in the research literature, there are few texts that discuss transdisciplinary community engagement. Therefore, the following section looks at concepts of community and community engagement models in relation to case studies presented by contributing authors through the lens of transdisciplinarity, in order to expand our understanding of this emerging concept.

Community as an Idea

Community, as debated in Chap. 1, is a broadly defined concept that is understood through socio-cultural-political-economic theories and practice, and therefore continually reconceptualized.

When the editors speak of community engagement, we are talking about relationships, an attitude of practice, a way of acting, mutual respect. This is where community always has the potential to be an equal partner. We recognize there are times and contexts where this way of operating is not appropriate (Guggleberger and Dur 2011; Roberts 2000); for example, when there are disastrous events, lives are lost or damaged and pandemic disease may break out, and urgent action is crucial. In this case authoritative, expert–led action is likely to be the most expedient and relevant.

However, as we were looking for collaborative models, the authors invited to contribute to this book were those working in community engagement projects that did not involve emergency and disaster relief. Therefore all projects in the book have benefited in some degree from a collaborative planning/action approach, and our hypothesis is that a considered, conscious transdisciplinary approach has even greater potential to build capacity by seizing possibilities and overcoming barriers to effective community engagement.

In this context, all things being equal, we maintain that 'the community', the site where the problem/action occurs, needs to be an equal partner in the identification of issues, development, action and evaluation. We maintain that others engaging must also be recognized as partners, with issues and needs of their own.

Within this transdisciplinary framework, we examine how the authors working across the fields of health, environment, ecology, social sciences, humanities, planning and design, have thought about community and the ways in which they interact with these communities.

In Chap. 1, the editors discussed others' conceptions of community engagement and their own positions. In previous work on community engagement, the editors conceptualized that community engagement is 'about learning and exchanging knowledge, identifying priorities, possibilities, making decisions, and making things happen' (Beeck at al. 2011, p. 17), and is a 'cooperative process of working with people to address their wellbeing, crossing disciplinary boundaries, and using multiple knowledge from inside and outside the community' (Lommerse 2011, p. 26).

When the editors envisioned the shape and contents of this book, they hypothesized a framework and a diagram and called for abstracts that spoke to the model of community engagement as proposed in Chap. 1 (see Fig. 1.1). Each author/group of authors was then asked to map their case study onto the community engagement framework. From the individual case studies described in chaps. 2–10, we argue, a unique model was created in response to the nature of the communities that were engaged, the scale and type of project, and the disciplinary perspectives of the researchers. For a visual narration of the nine models, see Figs. 11.1–11.9.

To understand what constitutes a community, we now examine how the contributing authors have (a) conceived, and (b) engaged with a community.

a) Conceptions of Community

The contributing authors describe the community projects they have engaged with and report on by using various conceptions of community. Some authors describe the community as a group specific to a locality (Chap. 3, Singh et al.); others describe educational projects that conceive of communities as a group that shares common knowledge (Chap. 9, Farren and Spanbroek). Some describe practice-based projects that bring a group of people together in communities of practice (Chap. 8, McGann and Milech). Methods of engagement sometimes determine the conception of communities; for example, a 'digital community' built around a shared interest area, as in Chap. 7 by Deluca, is a community that is not place-based and is widely dispersed. In a number of chapters, the project profiled integrated several of these conceptions of community.

While the authors' case studies were each embedded in a specific community with its own issues, problems, strengths, knowledge and skills, the core characteristics of community have been identified through the chapters, and the methodologies and methods could be transferred to another context. Therefore the next section provides a glimpse of the concepts of community, the engagement between the communities, and how that may affect capacity building within and between those communities.

Example model one

Intent, aim and likely outcome McGann and Milech's model (Chap. 8) offers pro-bono architectural design services to communities in a way 'that facilitates capacity-building opportunities for three cohorts: academic researchers, early-career graduates and industry professionals through a collaborative framework.' The fourth participant, the client/user, gained in the capacity-development effect of the project. The tangible outcome was the production of a twin-house design for a disabled person.

Characteristics The model sits in a research and praxis nexus in architecture and design, in a suburban neighborhood, in a developed country. The architect/researcher/practice/client engaged through using the tools of design thinking and creative practice, Implicit in the engagement was the valuing of the client/user's experience and knowledge in the process of development, decisions and actions (see Fig. 11.7).

Capacity building McGann and Milech report a considerable gain in capacity. In their conclusion, they describe the following outcomes. 'The client gained a design far beyond his expectations and developed ownership of the project through the consultation stages.' Importantly, he gained hope and a vision for his future. 'The graduate/recent employee gained strong client/community experience, formed a bond with the client and grew to understand the everyday diverse needs of wheelchair users'. The academic developed the capacity to publish, and the professional practice gained the capacity to develop an employee's skills. 'Each participant in this project gained something pertinent to their life circumstances and professional trajectory.'

Future potential the innovative twin houses design for a disabled person has evidently brought transformative and positive change to the life of a person with disabilities, a community of one, which expands to include family, carers and neighbours. This community is a special sector of society, dealing with physical capacities that are not the 'norm' in society; therefore the 'normal' built environment (inside/outside, domestic/public) and the constructs/policies that determine them do not adequately support the wellbeing and livelihood of this sector. The support structures are so silo-thinking that innovative solutions such as this are unusual. The overlapping interests of the two communities (one a disabled family, the second an academic/graduate/practice seeking development opportunities) meet and mesh rhizomatically to create a hybrid—an innovation, perhaps a transdisciplinary hotspot. The model can be usefully applied to other contexts. For example, for the wider disabled community in a suburb, city or region, creative practitioners can work with the residents, and with social service specialists and/or policy makers to create a more accessible and livable habitus, which would illustrate to the neighborhood and beyond that a person with disabilities can lead a meaningful and productive life in the community.

Example model two

Intent, aim and likely outcome In Farren and Spanbroek's model (Chap. 9), the case study utilized a cross-sector approach, engaging businesses, council, educators, students and the private sector. The aim, in response to the changing needs identified by the city council, was

> …to enrich the existing social, cultural and business environment. Their vision was to create a …[city centre] that was cohesive and coherent and that retained the unique sense of place of each of its public spaces. [The tangible outcome was the visual material developed, which provides] a database of ideas …that enabled the local community to see their city through fresh eyes.

Characteristics The model uses social engagement and street activation as a catalyst for change. It is sited in an inner urban centre, in a developed country. Design students, creative practitioners and academics engage in a re-examination of the city, developing ideas which presented a new vision for public sites, with a focus on issues of accessibility and liveability. They actively engaged the local community in dialogue and interaction with the '"re-visioning" process and re-examination of ideas about the city.' The core characteristics are revealed through the parties who engage in creative change. The outsiders share values around design practice, design thinking, learning and knowledge development, people-environment relationships and creative practice. The insiders hold and open up for creative engagement the culture of their city, its livability and vitality, its unique heritage and attributes of place (see Fig. 11.8).

Barriers and challenges identified Time for planning and development and finite funding limited the potential for a project of this scale.

Capacity building Farren and Spanbroek report that the students gained capacity, strengthening their learning of design, technical skills and ability to engage in real-life situations and collaborative work. They state in their conclusion, 'There is recognition of the economic value and significance of cultural…[development].' Therefore, this indicates that there is community value in the projects that 'engage the community in reflection on and identification of its unique status' suggesting the model has potential for building the community's capacity.

Future potential Farren and Spanbroek's framework makes rhizomatic connections with: issues defined by community of locale: council, people on the street, store owners; and students and their teachers and mentors. The 'inside' and 'outside' communities come together in a creative assemblage—the exhibition of the findings and possibilities. The harnessing of the objectives of two parties through a problem to be resolved opens new understandings, new creative projects, new solutions to community needs. The model will be valuable for communities interested in building their community through the recognition and valuing of local culture and design institutions wishing to expand knowledge of community engagement.

Example model three

Intent, aim and likely outcome The third example, Chap. 4 by Mendoza-Arroyo and Vall-Casas, is very different to the prior two, as it examines methodologies and models used, and the built public spaces that result. The tangible outcome is the analysis of the urban renewal tools in terms of the 'relationship between the participatory process and the urban design quality.'

Characteristics The model assesses participatory and inclusive decision-making in urban neighbourhood regeneration, and is situated in the disciplines of urban planning, urban design and social sciences. The case study is sited across a large metropolitan region in a developed country. In this model the authors examine the participatory processes that involved consultation with hundreds of citizen participants, dozens of districts, their councils and the planners, and that resulted in extensive regenerative planning and building over a 20 year period (see Fig. 11.3).

Capacity building The capacity built in this model is knowledge—important knowledge about common methods, and theory about participative processes and how they work/don't work in reality.

Future potential The model provides a framework for other disciplines investigating participatory practices, and other regions wishing to explore similar issues identified in the case study.

It is one thing to define a community, but it is another to engage with them. The following point talks about additional models, and the authors' different engagements with community.

b) How are Communities Engaged?

We now foreground how the contributing authors have engaged with the communities.

Example model four

Intent, aim and likely outcome Chapter 2 by Tiwari and Pandya describes a community action project for a slum community, where different stakeholders had different objectives for the project. The tangible outcomes were 'a less polluting environment, as a result of recycling of municipal waste…[and] economic benefits to the community through the construction of affordable built forms' such as an activity centre—both the physical space and its program of activities and services.

Characteristics The model sits in an ethnographic and collaborative model of inquiry, using creative practice in the discipline of architecture. The case study was situated in a lower socio-economic community (a slum) in a developing country. The authors placed importance on partnership, and the building of capacity in each of three partners collaborating on the project: the community of locale where the action is taking place, the NGO, and the students and researchers. The tools of engagement were: relationship building with stakeholders, observation, participant observation, scenario building, co-design workshops, informal chats, unstructured interviews, co-construction, awareness programs and community events.

> *Capacity building* …strategies include providing residents with skills to participate and identify their needs and issues, encouraging them to express their views on local issues before co-designing and co-constructing with them, letting them make informed decisions about their priorities, and establishing a wide range of participation and representation structures. All these are effective in encouraging reciprocity, skill development and innovation.

(See Fig. 11.1)

Barriers and challenges identified Tiwari and Pandya stated that

> … the aim of collaborative planning was to have a horizontal relationship between all stakeholders, with everyone being equal to each other. However, the challenge was that on the ground things were different. All efforts to make this project successful were channelled through the NGO….

and thus, power resided with the NGO and resulted in the role of local government being negligible. The authors consider that it is essential to involve local government bodies and other legislative bodies so that the planning and financial measures can be re-oriented for a greater involvement of residents in activities that meet their needs and aspirations.

Capacity building Tiwari and Pandya report that capacity was built for the NGO, community, local practitioners, architecture students, and researchers, and that: 'Reciprocity, learning and creative innovation are the most significant capacity building outcomes (Innes and Booher 2003) and these are reflected tangibly and intangibly in the Activity Centre initiative.'

Future potential The roles played by different stakeholders (slum dwellers, NGO, students and researchers, practitioner) could be conceptualized as rhizomatic and holistic. The model is inclusive in that it provides the opportunity to have a say to the socio-economic groups who are generally left voiceless in the negotiations and decisions about their right to city spaces. This could be a vehicle for similar work in the future.

Example model five

Intent, aim and likely outcome In Chap. 10, Di Mauro uses co-creation, engaging the local community to solve an urban planning problem. The tangible outcomes were concepts for the 'Ideal Town' project. However it is the intangible outcomes, brought about through the process whereby the author engages with the community, that he underlines.

Characteristics The model sits within public art and design practice, and is designer/artist-directed, using community-led, sustainable practices. The case study is sited in the rural and urban regions of a developed country. Di Mauro use tools such as: archival research, exhibition/showcase, display, sounding board, the act of disclosure, workable templates, progressive evaluations, and hands-on activity.

Di Mauro describes how, when working with community as a partner, he considers how to develop the capacity of the community in regard to developing ideas and actions. He describes how he engages the community in the contribution to knowledge, and expands on the methodology he uses to co-create:

> ...the artist becomes the catalyst for the sharing of local knowledge. The artist moves from a process of being solely responsible for injecting an expressive content into the community to one of the community being proactive in the contribution of knowledge towards an unknown end.

(See Fig. 11.9).

Capacity building Di Mauro reports that capacity was built for the community, in terms of understanding the creative process, confidence building through transparency, equity and inclusion, breaking down barriers and negating difference. He describes his methods for building the recognition of skills and ongoing economic capacity in a community, working in partnership to develop the capacity of the community through creative ideas and actions:

> In an effort to extend on notions of ownership, it is important for the artist to seek out various skills that exist within the community or be prepared to train others to build up their skills in order to keep local content at its maximum. Quite often this retraining or reawakening of dormant skills continues to energize community and local industry long after the project has been completed.

Future potential The model reveals the transition of disciplinary practices and values (of artists) into a community engaged model. The value of this model is the ways in which a discipline can transition. This relates to some of the reflective practice, the 'aha' moments in Chap. 12 by Smith:

> ... for those entering community work, either as an academic or as a discipline expert, there is a need to consider what it is to be a novice in this new type of domain... in the professional development of a community engagement practitioner.

The model itself has the potential for application across community development, planning and design practice engaged with community. It also could be trialled in areas such as health.

> Creative art and design pursuits, especially if deeply reliant on an understanding of place, provide an ideal opportunity to identify and showcase knowledge and skills which exist within the local community but which quite often are never made public. Through such showcasing of events, employment and commissioning opportunities for artists and crafts workers are realized, and may often lead to new financial avenues for the production of new art and cottage industries.... The opportunity exists for the artist to showcase and encourage local knowledge and skills, and in the process provide sub groups with an opportunity to collaborate and contribute to their community.

Example model six

Intent, aim and likely outcome Chapter 6 by Reid et al. describes the model involved in a research collaboration that was established between the Center for Rural Development (CRD) (India) and the Centre for International Health (CIH) in Australia. The initial requirement of CRD was ‘an assessment to gain a better understanding of available health care facilities and access to the health of local communities’ in a district within a developing country. The tangible outcome is a practical approach to applying participatory action research (PAR) in communities, from assessment of community needs, to an example of addressing the identified needs in a case study.

Characteristics The model sits in PAR methodology, which places the power of problem identification and decisions on action in the hands of the affected community. Situated in community and international health, the project assesses and addresses community health needs. The case study is situated in tribal communities (Indigenous disadvantaged peoples) in five rural villages in a developing country. The tools used were: participatory action research methodology; participatory approaches; rapid participatory appraisal (RPA) observation; focus groups and interviews with key informants; secondary data/documents and community meetings (see Fig. 11.5).

Barriers and challenges identified Read et al. identified a number of barriers that related to contextual factors in the research environment. Power relations within the community, the researchers’ and community members’ time limitations, and organizational capacity, all had a role in affecting the transformation of knowledge to action.

Capacity building Reid et al. report that the capacity built includes learning, social engagement, raising awareness of community health issues, and encouraging communities to realize their own potential in addressing community needs.

Future potential Read et al. present a model that can effectively guide community-based action researchers working in the area of community health ‘to produce knowledge, action and change, whilst empowering local people to construct and use their own knowledge in order to emancipate themselves from their situations.’ Being context-specific, participatory approaches to research can be applied to any study with its own methodology. Therefore the potential is for this model to be trialled and/or modified by many other disciplines to see what emerges in a transdisciplinary context.

The authors in examples 4–6 set up their models with the explicit intention of having the affected community participate in all aspects of the process, including problem definition, consensus on priorities, identifying and securing resources, developing strategies, actioning, and post-project reviews. It is worth noting that Tiwari and Pandya explicitly include the other stakeholders (academics/students’ community of learning and knowledge and the NGO) as having equal partnership in these aspects, whereas the other two sets of authors do not mention this. We suggest that this awareness of the other stakeholders is a connection in the transdisciplinary, rhizomatic assemblage that can help to ease difficulties and meet challenges as they arise.

We see, then, that six different sets of authors conceive of community engagement and partnerships in different ways, depending on their disciplinary views, as well as the context within which they are working. This points to the relevance of the transdisciplinary model, which will be appropriate to the time, place and resources available. Many of the authors have adopted a core principle of 'community as equal partner', emphasizing the importance of knowledge, skills, values and social capital for community capacity building.

Building on the diverse conceptions of communities and ways of engaging, the next section explores some of the barriers, challenges and potentials uncovered in the models.

Barriers, Challenges and Possibilities Within Community Engagement

We wish to open up the debate about other ways of knowing and participating in community-based projects. But to do so, we first need to consider the numerous barriers and challenges that may arise during implementation of community engaged projects, ones that can turn action into an idealistic pursuit rather than a reality. Many of these issues relate to the systems that are embedded in social constructs, and are embedded in ways of working and disciplinary values, beliefs and practices.

Understanding the contextual factors that may impede the ability to influence positive change in communities can help to better direct limited resources.

Many of the community-based projects that are described in this book are multidimensional and undefined, that is, they involve 'wicked' problems. Locational aspects—whether the context was a developing or developed country, or a rural or urban setting—contributed to specific and different barriers and challenges. In Chap. 11, the editors examined the key barriers and challenges that emerged: (a) power relationships, (b) time and timing, and (c) finances; here we elaborate on these.

a) Power Relationships

Power relationships are implied in all the contributed chapters in this book. The examples are diverse, complex, and context specific; examples include:

- Researchers, as outsiders to the community, emphasized the need to respect the capacity of the 'insiders' in the situation to make decisions. For example, a two-(or more) way knowledge-sharing framework was set up instead of a commonly held perception that the 'outsider' carries expert knowledge (Chap. 2, Tiwari and Pandya; Chap. 8, McGann and Milech; Chap. 10, Di Mauro).
- Cultural protocols differ from context to context, and gender-based barriers were strong, preventing some women from joining groups, seeking care, or from implementing strategies if they had joined a group (Chap. 3, Singh et al.; Chap. 6, Reid et al.; Chap. 14, Reid).

- Issues of power imbalance (within the community) and insider-outsider tensions related to privilege and control affect the engagement and outcomes:

> Distribution of information, time, formal education and income reflect broader social inequities structured around race/ethnicity, class and gender, and these affect who participates, whose opinions are considered valid, and who has influence over decisions made in a community based research partnership. (Chap. 13, Rahman and Pokrant)

- There may be an imbalance of expectations of each partner in terms of their involvement and the level of action and change expected from the project (Chap. 2, Tiwari and Pandya; Chap. 9, Farren and Spanbroek; Chap. 12, Smith; Chap. 14, Reid). For example, the expectations of community partners' participation and/or community members' available time and motivation may be overestimated. Opinions about the appropriate timing to implement the interventions may differ substantially between partners.
- Project leaders and their support teams are themselves entangled in local power relations and struggles rather than being mere technical service providers. Local priorities often differ from project or state priorities, and researchers must be ready to re-think project directions in the light of experience. Giving local people a stake in project work is essential to project implementation, and researchers must be ready to see project objectives change, both as a result of increasing knowledge about the local circumstances, and also through the influence of local people wanting changes to how and for whom the project is being carried out (Chap. 13, Rahman and Pokrant).

These power relationships impact heavily on the other two factors summarized below: time and timing, and finances.

b) Time and Timing

Time was a limiting factor that all the case studies shared. For example, the time limits of researchers working within the academic calendar often placed restrictions on what could be accomplished, as did the time frame of each project itself as defined, for instance, by the funding body. The longevity of a project, constrained by these limits, affects the effectiveness of capacity building. Time and timing also needs to be considered within the affected community.

> …whilst it may seem logical that seasons would be accounted for, it is not uncommon to find that little thought goes into the timing and duration of participatory activities (Cornwall 2008). This also includes requesting involvement of people who work, people who have small children to put to bed or feed, people who are unable to justify spending hours outside the household (Cornwall 2008; cited in Chap. 14, Read).

The time-consuming nature of collaborative community-based projects is well-known. The more stakeholders, the more people with whom to communicate and get agreement. In addition, the need for coordination and communication, and to build the skills of collaboration, participation and team-based approaches to problem-solving and decision-making, take additional time and resources.

c) Finances

Finances constrict change agents' ability to commit to an ongoing program, and are usually required to build the capacity of community-based initiatives. For example, a number of the case studies illustrated the first phase of an action-based cycle, where the various players had got to the stage of problem identification, and developed the relationships around change and agreement about the way forward (Chap. 3, Singh et al.; Chap. 5, Madigan and Morris; Chap. 6, Read et al.; Chap. 12, Smith; Chap. 14, Read).

Additionally, finances constrict the affected community members' ability to commit to a project. Often they find it difficult to take time away from other commitments such as family duties and income generating activities. Access to finances can also support the development of income-generating activities for training, equipment and infrastructure such as micro-financing of start-up operations.

Finances for a project can create potential. They can break poverty cycles, for example, by supporting local industries to buy required infrastructure, materials and/or business expertise. And they can increase health funding for vaccines, health programs, medical education and services.

Common Ground for Connections

It is recognized that taking community action with active involvement of communities is challenging, and contextual challenges can effectively halt any action to overcome community issues and address problems. Therefore, Part II of the book explores the common ground between the different disciplinary approaches and offers reflections and tools for consideration. In Chap. 12, Smith provides examples of the barriers that may arise during community projects, ranging in scale from micro-interior projects to larger village and town settings. A valuable tool is the 'hints for practitioners' (Table 12.3), which offers suggestions for community engagement practitioners who embark on community-based projects. In Chap. 13, Rahman and Pokrant relate experiences over 20 years in several action research and academic studies of natural resource management in Bangladesh. The authors explore the relationship between communities of practice and communities of place. Rahman and Pokrant state that putting local people first can range from local people being co-producers of the project design, implementation and evaluation, to being simply beneficiaries or 'objects of study'. All the projects described in Chap. 13 fall between these two poles in that they sought to increase local participation in project activities. The content of this chapter provides a number of reflections which can assist others directly, as well as highlighting the complexity and the evolution of place-based projects of this scale.

Chapter 14 by Read presents a new tool, formulated as an instrument to identify and categorize context-specific challenges. It is intended to help stakeholders identify potential road blocks in their project, which will give them options to develop proactive strategies. Read introduces an Action Preparedness Tool for Community-

Based Projects, designed for the stage between identifying community issues to be addressed and taking action to address the issues. This tool will help stakeholders identify potential challenges *before* they take action. Chapters. 12–14 each explore power relationships from their disciplinary perspectives.

The collaboration between disciplines, in which we include the community, has its challenges and barriers. For the researcher, the challenge is to detach themselves from their discipline, in particular their traditional research methods (Ramadier 2004), without completely disregarding their knowledge (Hoffmann-Riem et al. 2008). As Russell et al. phrase it, the thought of the researcher can often be 'where is the place for people in our knowledge?' (Russell et al. 2008, p. 464). The next section will address this challenge by suggesting that conceptualizing community as a discipline will overcome artificial, hierarchical barriers.

Community as a Discipline

Approaching the community or public as a discipline is a necessary value in transdisciplinary methodology. For any actor truly trying to approach social problems in which the community is the key stakeholder and the end-user, the community must be identified as having equal status to all others.

To begin, it is useful to restate the editors' positioning of how community operates as its own discipline from Chap. 1. This sets the groundwork for a discussion on what emerges from the contributing authors' chapters. The editors explored a reconceptualization of community as a discipline, in order to embed the equity of the players in a participatory project that stresses community engagement; that is, one where the community, gifted with expert inside knowledge, invites external experts to work with them and share their outside knowledge, in order to understand and address locally defined issues. The community members are participants who are both the hosts and meaningful contributors. Chapter 1 outlines this premise with the recognition of a number of attributes of transdisciplinary community action for change:

- the community owns and/or belongs to the location of the project;
- the project activities will occur on community members' territory;
- the project will disrupt the community's existence in some way;
- the consequences (positive and negative) will remain with the community after the life of the project (Chap. 1, Smith et al.).

The concept of community as a discipline implies a framework for action which avoids authoritarian intervention. A core principle of this discipline is a communal approach to change that engages a collective of people, insiders and outsiders, with unique and embedded practices and understandings that they bring to the project as it is explored, understood, and pursued to bring positive change. This principle is informed by the recognition of outside experts, who may be from various disciplinary backgrounds, of a community's depth of tradition and fluidity. The outsiders accept the uniqueness of the community and its ability to fully contribute in unforeseen ways. The transdisciplinary agenda of community as a discipline seeks new understandings by bringing the unlike together and interrogating the edges between the two. This shift opens up greater, perhaps previously unthought of opportunities, when multiple players, including the community, are involved.

In the next section we look at three more case studies, and explore how the community engagement framework we see at work here, and the models that emerge, have potential in other contexts.

Example model seven

Intent, aim and likely outcome In Chap. 5, Madigan and Morris's project 'arose out of the collective aspiration of the Tannese people to develop a community-managed tourism enterprise based on a sustainable business model for developing a trust fund for education.' To date, the first phase of what is expected to be a 5–10 year process has occurred. The tangible outcome is the 'big idea', the design with drawings and models of a visitors' centre.

Characteristics The model uses hybrid architecture and creative practice and sits across the disciplines of architecture, civil and environmental engineering, tourism and hospitality management, business management and health and education management. Working within an indigenous lower socio-economic community, it is sited on a rural, remote island, in a developing country (see Fig. 11.4). The tools used to date are: experiential immersion in the context of the project, community consultation, site survey, architectural and engineering design and communications, public exhibition.

There is evidence that the community of Vanuatu (and the island of Tannu in particular) are acting as a discipline. The Tannese people identified their issues and developed a potential solution: to develop a community-managed tourism enterprise that would become a sustainable business model for developing an education trust. They sought expertise not available in their community to work with them towards a solution. The Tannu community are participants who act as the hosts and meaningful contributors and meet the attributes of a discipline; they have a strength in their explicit and tacit values, beliefs, and bring their own knowledge and skills to the project. They used their own rituals, engagement and practices to immerse their partners/other discipline (architecture) in the culture of the place.

> Our arrival in the community was marked by two unexpected and powerfully affecting events…the combined acts of hospitality and inclusion gave us opportunities to be immersed in village life in a meaningful way, as well as a deeper understanding of the project's requirements. (Chap. 5, Madigan and Morris)

Capacity building Madigan and Morris report that the intended long-term capacity building on this project includes: cultural sustainability; cross-cultural awareness; economic independence; socio/political engagement in a process of change wherein community members open up their habitus for tourism; and education of participants (outsiders and insiders) in ways to build environmentally sustainable and culturally appropriate systems.

In this first phase, the authors report that capacity built within the community is the ability to witness and understand formal planning processes, design and spatial conceptualization, and strategic thinking around greater project implications. For the students, capacity is built by placing them in the roles of professional practitioners working with real clients on a tangible project, across cultures and in the field, and away from their familiar office-based design tools.

Future potential Vanuatu displays the attributes of a 'wicked problem' including: minimal access to basic health and education, poor infrastructure, and negative impacts from economic development, such as cultural disruption caused by the incremental transfer of indigenous land ownership to commercial interests from outside Vanuatu. Here there is potential for transdisciplinarity to address the following challenges: how does a community who has identified a solution to the identified problem of funding education for their children by erecting a building to attract western tourists (seemingly without the skills and knowledge needed to design, build and then manage such a thing) move forward at a reasonable rate of development? And how do academics and students (a community of knowledge) wishing to engage with such a community, but having the restraints of a limited time in place (2 weeks in Vanuatu in this case), engage with the community to realize their goals in a meaningful way?

Madigan and Morris describe a traditional client/architect methodology and use it as a model for the way to work. They have listened to the client, investigated the context and presented a solution for comment. At this stage, only two disciplines are actively involved—the Tannese community and architecture. The project partners anticipate that the project will require further work with the community to build capacity to organize, manage, and maintain the project. The chapter however is focused solely on the building project and the running of it, which is the realm of the external discipline—architecture—that has been brought in to date. It is possible that a transdisciplinary approach would benefit other aspects of the 'wicked problems' facing the Tannese community.

There is a sense that there are unmined strengths and possibilities within the community. With outside collaboration, these could tap into new opportunities and models of tourism (authentic experiences, eco-tourism) thus building capacity while protecting the fragile environment and disappearing cultural ways. There is exciting potential here for a truly innovative solution that can act as a pilot for further development in Vanuatu and other remote and sensitive environments, achieving transformation.

Example model eight

Intent, aim and likely outcome In Chap. 3 by Singh et al., the aim is to develop an environment that supports community health. This case-study illustrates the initial steps in the long-term process of addressing barriers to the full inclusion of women in decision-making processes within rural communities. A tangible outcome is that through initial dialogue, the community identified a number of challenges and opportunities which could be addressed.

Characteristics The model sits across the disciplines of international health, international development and rural and maternal health. The case study involves men and women of lower socio-economic status and is sited in several rural villages in a developing country.

> Community-led capacity building applies and uses insights drawn from the community to address community concerns. Consciousness, reflection and sharing are critical steps in capacity building. The process of planning, action and reflection undertaken in this study ensured that an increasing depth of understanding was achieved by the participants. (Chap. 3, Singh et al.)

This case study was conducted in a post-conflict, transitional, developing nation; twenty-one men and women in central Uganda participated in interviews and focus group discussions. Tools used were: participatory empowerment approaches; focus groups; interviews with key informants; community dialogue; and use of cultural interpreters (see Fig. 11.2).

Capacity building Singh et al. report that capacity built through this long term project is and will be: awareness of community maternal health issues; encouraging communities to realize their own potential in capacity building and community development; and a community dialogue on the need for greater equality of genders. As this was an early phase of a long process, early lessons were learnt on the relationship between gender-related decision-making, empowerment and meaningful participation and how these inter-related concepts contribute to community development.

Future potential the model has the general potential to be applied to any discipline.

Example model nine

Intent, aim and likely outcome In Chap. 7, Deluca states that:

> [b]uilding the capacity of communities to achieve sustainable design requires developing community leaders who can choreograph diverse teams of creative, idealistic people able to realize radical innovation in a complex and rapidly changing world. A pilot online course called Creative Leadership was taught with the goal of empowering sustainable design students to become effective leaders.

The tangible outcomes of this case study are early indications by students of their preparedness to lead in complex and uncertain situations.

Characteristics The model sits across the disciplines of civil and environmental engineering, biomimetics, biomimicry, sustainability and sustainable design. It is a teaching and learning model sited in an online environment, engaging students mostly in the developed world. Sustainable design requires radical innovation.

> Conventional leadership is very good at driving incremental innovation, but is ill suited to drive and deliver radical innovation. Conventional leadership is also inadequate in increasingly interconnected, complex, unpredictable, and volatile conditions. Creative Leadership uses nature as a model for the leadership principles and practices, because evolution has yielded organisms that are responsive, adaptable, resilient, able to leverage and support dynamic complex systems, and embrace unpredictability and disturbance. A Creative Leader continually maintains and refreshes the common vision, values and goals, sensing

and adapting to changing conditions both inside and outside the community, and creating conditions of trust, respect, and curiosity that enable dynamic synergisms and co-creativity. (Chap. 7, DeLuca; see Fig. 11.6)

Capacity building Deluca reports that capacity is built in intellectual knowledge, emotional understanding, and physical skills. 'Initial feedback from students suggests a shift in thinking and increased capacity to lead in creative, idealistic, and challenging situations.'

Future potential We observe that this model potentially prepares future leaders to be ready to look at things in new ways, prepared for 'wicked problems'. This model would be appropriate for training agents to rethink assumptions, and is highly suited to transdisciplinary situations. A number of the models and some of the literature studies cite the need for co-ordination, leadership, management and coordination suitable for complex projects with undefined issues, and fluid environments.

When community emerges as a discipline that replaces the more traditional, hierarchical subject-object or 'us and them' relationship of community intervention with a relationship of equality, openness, respect, listening and flexibility, the outcomes are potentially transformative for all involved in the process. This next section explores this idea.

Capacity Building: A Transformative Process

Capacity building, as defined in Chap. 1, is about 'empowering the community members involved, the researchers facilitating the project and/or the organizations to which they belong.' Overall, a transdisciplinary perspective opens up opportunities for new ways of conceptualizing community. Power-sharing, and a shift in decision-making to include those within the community who are affected during community engagement, are important for a successful capacity building outcome. Chaskin (2001), from a comprehensive literature review on community capacity building, came to the following definition:

> Community capacity is the interaction of human capital, organizational resources, and social capital existing within a given community that can be leveraged to solve collective problems and improve or maintain the well-being of a given community. It may operate through informal social processes and/or organized effort. (p. 295)

Much of the literature suggests that whilst every project context is different, community capacity building is inherently process oriented, and the process is much the same across contexts (Mathie and Cunningham 2003; Chaskin 2001; OECD 2009; Suwanbamrung 2010; Merino and Carmenado 2012), although different disciplines have different perspectives on their positions and roles.

It is our contention that community engagement and capacity building can and should be used across socio-economic contexts; the models in this book provide examples of a range of socio-economic contexts where capacity can be built. As

Singh et al. Chap. 3 so aptly put it '[t]here is no community that can claim to be 'developed'. All communities are on a continuum of development, with its inherent challenges.' In development work, the nature of the projects inherently address social minorities and the economically marginalized (OECD 2009; Mathie and Cunningham 2003). Many disciplines view the use of community capacity building as a tool for empowering disadvantaged communities. Many of the rising community arts engagement projects in Australia are dealing with disadvantaged communities (Ang et al. 2011). Also, among the general public across socio-economic groups, arts and culture are becoming recognized for making communities more liveable and enhancing the lives of all. Chaskin (2001) states that the level of commitment required for effective community capacity building means having residents that see themselves as stakeholders in the community and are willing to engage. Furthermore, communities with abundant resources (higher socioeconomic demographic) are more likely to see themselves as stakeholders, and have political influence (Chaskin 2001; Mathie and Cunningham 2003; OECD 2009). Regardless of this challenge, many disciplines see the use of community capacity building as an effective process for engaging social minorities and communities at the lower end of the socio-economic scale (Mathie and Cunningham 2003).

Editors considered: what is the effect of transdisciplinarity on the individual discipline? When the discipline's community of learning and knowledge is involved, there are outcomes for the discipline such as new knowledge, new methodologies and extensions of practice, 'because the actors develop new capacities… [and] an ability to frame and contextualize sources of uncertainty and conflict' (Fontan et al. 2013, p. 312).

From a disciplinary point of view, capacity building, we argue, should be an explicit aim in community engagement projects (Chaskin 2001; OECD 2009), but its value and position as a key project aim can differ across disciplines and projects within specific disciplines.

In development, health promotion, planning, ethnography, and other disciplines dealing with public problems, the agents see themselves as a catalyst for transformation in a system or process (Barab et al. 2004; Mathie and Cunningham 2003); that system or process is central to the project's aims, making it explicit in nature.

However, there are projects where capacity building is implicit, such as community arts projects that are ongoing, establishing networks and partnerships, and where capacity building is just a part of the whole process (Ang et al. 2011).

We suggest that, from the community's point of view, whilst community capacity building should be foregrounded in project development to effectively achieve the desired outcomes, which will primarily be that of sustainable community development (Mathie and Cunningham 2003; OECD 2009), the agent must step back to make room for the community's own empowerment and self-determination (OECD 2009). To this extent, then, the aim to build capacity is implicit, a grounding principle of community engagement, and an expression of community's desire to better itself.

Conclusions

We (the editors) gained three key insights through the process of developing this book: the applicability of community engagement across all strata of community; a need for development within and between disciplines; and equality.

Firstly, the contributions have shown that community engagement can be extended beyond the customary interpretations within the various disciplinary constructs. In doing so, the benefits can expand knowledge and action for all the communities involved in community-based projects, and pave the way for improved future collaborations. This transdisciplinary model promotes greater synchronicity of knowledge production and transfer, more accessible and relevant knowledge, the acceleration of results, and a fuller exposure of research outcomes to the community and the academy. The methodology combines and, most importantly, *builds* the skills sets required for co-producing knowledge (Fontan et al. 2013).

Secondly, there is a need for disciplines to develop models for engagement of the various stakeholders and social actors. A shift from traditional discipline practices into community-engaged project work requires a more inclusive and extended field of expertise, as does the ability to work in collaborative, co-creative environments across disciplinary boundaries, meeting the challenges of complex, ill-defined issues, and moving targets.

By exploring the diversity of contexts offered by contributing authors, our instinctive modes of operating were challenged when we considered the professional development of a community engagement practitioner. We need to be cognizant of our biases and assumptions, and develop tools for planning, management, communication and knowledge dissemination in complex contexts, with the objective of facilitating individual and collective growth.

Thirdly, there is the matter of equality—giving voice to all different kinds of people. We maintain that 'the community', the site where the problem/action occurs, needs to be an equal partner in the identification of issues, development, action and evaluation; and that others engaging must also be recognized as partners, with issues and needs of their own. Awareness of the other stakeholders creates powerful connections in the transdisciplinary, rhizomatic assemblage that can help to ease difficulties and meet challenges as they arise.

Future work needs to be done to develop a more relevant definition of community, one that partners in collaboration can draw on to establish clearly *what* and *whose* capacity is to be developed.

A new concept of community has emerged from this book, which has explored community-based projects from different disciplinary areas through a transdisciplinary lens.

> Only those who will risk going too far can possibly find out how far one can go.
> T. S. Eliot, preface, *Transit of Venus*

Acknowledgements Many thanks to Jessica Winters for contributing to the literature review.

References

Ang, I., Lally, K., Anderson, K, Mar, P., & Kelly, M. (2011). Introduction: What is the art of engagement? In E. Lally, I. Ang, & K. Anderson (Eds.), *The art of engagement: Culture, collaboration, innovation*. Crawley: UWA.

Baldwin, C. (2013). Living narratively: From theory to experience (and back again). *Narrative works: Issues, Investigations and Interventions, 3*(1), 98–117.

Barab, S., Thomas, M., Dodge, T., Squire, K., & Newell, M. (2004). Critical design ethnography: Designing for change. *Anthropology & Education Quarterly, 35*(2), 254–268.

Beeck, S., Smith, D., Lommerse, M., & Metcalfe, P. (2011). An introduction to social sustainability and interior architecture. In D. Smith, M. Lommerse, & P. Metcalfe (Eds.), *Life from the inside: Perspectives on social sustainability and interior architecture* (pp. 24–57). Perth: Pencil and Paper.

Cassinari, D., Hillier, J., Miciukiewicz, K., Novy, A., Habersack, S., MacCallum, D., et al. (2011). Transdisciplinary research in social polis. http://www.socialpolis.eu/uploads/tx_sp/Trans_final_web_double_page.pdf. Accessed 10 Oct 2013.

Chaskin, R. J. (2001). Building community capacity: A definitional framework and case studies from a comprehensive community initiative. *Urban Affairs Review, 36*, 291–323.

Conklin, J. (2006). *Dialogue mapping: Creating shared understanding of wicked problems*. Chichester: Wiley.

Conklin, J., Basadur, M., & VanPatter, G. K. (2007). Rethinking wicked problems: Unpacking paradigms, bridging universes (Part 1 of 2). *NextD Journal, 10*(1), 1–30. http://humantific.com/wp-content/uploads/2009/07/NextD.../NextD_10_1.pdf. Accessed 11 Oct 2013.

Doucet, I., & Janssens, N. (2011). Editorial: Transdisciplinarity, the hybridisation of knowledge production and space-related research. In I. Doucet & N. Janssens (Eds.), *Transdisciplinary knowledge production in architecture and urbanism: Towards hybrid modes of inquiry* (pp. 1–14). Dordrecht: Springer.

Deleuze, G., & Guattari, F. (1987). *A thousand plateaus: Capitalism and schizophrenia* (trans: B. Massumi). Minneapolis: University of Minnesota Press.

Fontan, J., Harrison, D., & Klein, J. (2013). Partnership-based research: Co-production of knowledge and contribution to social innovation. In F. Moulaert, D. MacCallum, A. Mehmood, & A. Hamdouch (Eds.), *The international handbook on social innovation: Collective action, social learning and transdisciplinary research* (pp. 308–320). Cheltenham: Edward Elgar.

Gugglberger, L., & Dür, W. (2011). Capacity building in and for health promoting schools: Results from a qualitative study. *Health Policy, 101*(1), 37–43. doi:10.1016/j.healthpol.2010.08.019.

Hoffmann-Riem, H., Biber-Klemm, S., Grossenbacher-Mansuy, W., Hirsch Hadorn, G., Joye, D., et al. (2008). Idea of the handbook. In H. Hoffmann-Riem, et al. (Eds.), *Handbook of transdisciplinary research* (pp. 3–17). Cham: Springer.

Leavy, P. (2011). *Essentials of transdisciplinary research: Using problem centred methodologies*. Walnut Creek: Left Coast.

Mathie, A., & Cunningham, G. (2003). From clients to citizens: Asset-based community development as a strategy for community-driven development. *Development in practice, 13*(5), 474–486.

Merino, S. S., & de los Ríos Carmenado, I. (2012). Capacity building in development projects. *Procedia—Social and Behavioral Sciences, 46*(0), 960–967. doi:10.1016/j.sbspro.2012.05.231.0150.

Miciukiewicz, K., Moulaert, F., Novy, A., Musterd, S., & Hillier, J. (2012). Introduction: Problematising urban social cohesion: A transdisciplinary endeavour. *Urban Studies, 49*(9), 1855–1872. doi:10.1177/0042098012444877.

Novy, A. (2012). 'Unequal diversity' as a knowledge alliance: An encounter of Paulo Freire's dialogical approach and transdisciplinarity. *Multicultural education & technology journal, 6*(3), 137–148.

O'Campo, P., Kirst, M., Schaefer-McDaniel, N., & Hwang, S.. (2011). Introducing a transdisciplinary approach to applied urban health research. In M. Kirst, N. Schaefer-McDaniel, S. Hwang, & P. O'Campo (Eds.), *Converging disciplines: A transdisciplinary research approach to urban health problems*. New York: Springer.

Organization for Economic Co-operation and Development. (2009). *Community capacity building: Creating a better future together*. www.oecd.org/australia/. Accessed October 10, 2013.

Ramadier, T. (2004). Transdisciplinarity and its challenges: The case of urban studies. *Futures, 36*(4), 423–439. doi:http://dx.doi.org/10.1016/j.futures.2003.10.009.

Rittel, H. W. J. (1988). The reasoning of designers. Working paper A-88-4 of the International Congress on Planning and Design Theory in Boston, August 1987. Stuttgart, Germany: Institut für Grundlagen der Planung, Universität Stuttgart. www.cc.gatech.edu/~ellendo/rittel/rittel-reasoning.pdf. Accessed 3 Oct 2013.

Rittel, H. W. J., & Webber, M. (1973). Dilemmas in a general theory of planning. *Policy Sciences, 4,* 155–69. (Reprinted in N. Cross (Ed.). (1984). *Developments in design methodology* (pp. 135–144). Chichester: Wiley).

Roberts, N. (2000). Wicked problems and network approaches to resolution. *The International Public Management Review, 1*(1), 1–19.

Russell, A. W., Wickson, F., & Carew, A. L. (2008). Transdisciplinarity: Context, contradictions and capacity. *Futures, 40*(5), 460–472. doi:http://dx.doi.org/10.1016/j.futures.2007.10.005.

Stolterman, E. (2008). The nature of design practice and implications for interaction design research. *International Journal of Design, 2,* 55–65.

Suwanbamrung, C. (2010). Community capacity for sustainable community-based dengue prevention and control: Domain, assessment tool and capacity building model. *Asian Pacific Journal of Tropical Medicine, 3*(6), 499–504. doi:10.1016/s1995-7645(10)60121-6.

Tress, B., Tress, G., & Fry, G. (2005). Defining concepts and the process of knowledge production in integrative research. In B. Tress, G. Tress, G. Fry, & P. Opdam (Eds.), *From landscape research to landscape planning: Aspects of integration, education and application* (pp. 13–26). Netherlands: Springer.

Index

R. Tiwari et al. (eds.), *M² Models and Methodologies for Community Engagement*,
DOI 10.1007/978-981-4585-11-8, © Springer Science+Business Media Singapore 2014

FSC
www.fsc.org
MIX
Papier aus verantwortungsvollen Quellen
Paper from responsible sources
FSC® C105338

If you have any concerns about our products,
you can contact us on
ProductSafety@springernature.com

In case Publisher is established outside the EU,
the EU authorized representative is:
Springer Nature Customer Service Center GmbH
Europaplatz 3, 69115 Heidelberg, Germany

Printed by Libri Plureos GmbH
in Hamburg, Germany